AN ENVIRONMENTAL
ODYSSEY

AN ENVIRONMENTAL ODYSSEY

People, Pollution, and Politics in the Life of a Practical Scientist

MERRIL EISENBUD

UNIVERSITY OF WASHINGTON PRESS
Seattle and Washington

LIBRARY OF CONGRESS CATALOGING-IN-PUBLICATION DATA

Eisenbud, Merril.
    Environmental odyssey : people, pollution, and politics in life of
a practical scientist / Merril Eisenbud.
        p.   cm.
    Includes bibliographical references.
    ISBN 0-295-96949-0 (alk. paper)
        1. Eisenbud, Merril.   2. Environmental engineers—United States—
Biography.   3. Industrial hygienists——United States—Biography.
I. Title.
TD140.E35A3   1990
628'.092--cd20
    [B]   89-70626
                                                                        CIP

The paper used in this publication meets the minimum requirements of the American National Standard for Information Sciences—Permanence of Paper for Printed Library Materials, ANSI Z39.48-1984.

∞

TO IRMA

with my appreciation and thanks
for many reasons
for many years

CONTENTS

PREFACE

ALTHOUGH THIS BOOK is autobiographical, it is far less about me as a person than the wide range of environmental problems with which I have been concerned for more than half a century.

Broadly defined, my activities have been involved with protection of human health from the effects of chemical and radioactive pollution, both in the workplace and the general environment. For the most part my work took me outside the laboratory to industrial factories and mines, to places where nuclear weapons were being tested, to the streets and the rooftops of cities, and to polluted rivers, lakes and bays. These were the places where environmental problems existed and where they had to be solved. I knew the importance of laboratory research and used it to assist and guide my work in the field but my main interests were in the world of noxious gases, radiations, dusts, and smokes. These are the unpleasant byproducts of life in a technological society and they must be studied so that their effects on health can be reduced or eliminated while at the same time retaining the benefits that technology brings to society. Public interest in such matters hardly existed when I began my career in 1936 and didn't begin to develop until the late 1960s when the environment suddenly emerged as a major focus of public and political concern.

This book had its origin in a lecture I delivered in 1985 at the Institute of Environmental Medicine of the New York University Medical Center on the occasion of my retirement to the position of professor emeritus. The lecture was an account of some of my professional experiences. It was addressed to the talented staff of the institute, many of whom were recognized worldwide for their scientific contributions to the subject of environmental medicine, but were so absorbed in highly specialized laboratory research that they were unaware of the excitement of having "hands on" experience with the problems that exist in the world outside the laboratory.

I enjoyed preparing and delivering that lecture so much and was so encouraged by the interest of my associates and others, that I decided to summarize my experiences in writing. This book is the result and is addressed to a variety of audiences.

To the many members of the general public who share an interest in the effects of environmental influences on health, and to concerned government officials, I hope this book will explain how ramified the subject is, and how important it is that priorities be properly aligned, and that difficult choices be made in an orderly manner.

To the many scientists, engineers, physicians, lawyers, and others who have entered the field of environmental health during the past two decades, I hope the experiences of one practitioner who began his work when the present legal, financial, and technical environmental protection infrastructure did not exist will be of interest. Finally, to young men and women who are choosing a career, this book will explain how exciting it is to work in the field of environmental health protection, and how much more needs to be done.

I wish it were possible to acknowledge fully my indebtedness to everyone who has been of help to me in the course of my career. Sheer numbers make it impractical to identify them all, although the names of many will appear in the chapters ahead. However, there have been a very small number toward whom I feel a special sense of gratitude. Among these during my earliest years were Professor Philip Drinker of the Harvard School of Public Health and Stuart Gurney, who directed industrial hygiene activities for the Liberty Mutual Insurance Company. For their wisdom, guidance, and counsel during my years with the United States Atomic Energy Commission, I am especially indebted to Bernard Wolf, Shields Warren, John Bugher, and my principal assistants in the Health and Safety Laboratory, William Harris, John Harley, Hanson Blatz, and Harris LeVine. At New York University I was fortunate to benefit from association with Norton Nelson and Arthur Upton, both giants in their fields who were successively the Directors of the Institute of Environmental Medicine. However, not all of those to whom I am indebted were scientists or engineers. There were also many devoted secretaries, administrative assistants, librarians, graphic artists, technicians, and graduate students. These and others in great numbers are deserving of my final expression of gratitude which, except in a few cases, has been much too long delayed and is too impersonally offered.

Many readers will understand my initial feeling of uncertainty at the prospect of having to reconstruct a sequence of events that took place over a period of more than fifty years. However, since much of my work has been reported in professional journals, the published papers have been an important resource. I have also been helped by my longstanding practice of maintaining diaries in which the entries, though often very abbreviated, were sufficient to remind me of people, places, and events.

I appreciate the assistance provided by Alfred J. Breslin and Edward P. Hardy, who reviewed the chapters that reported on events with which they are familiar. The entire manuscript was read by Leonard J. Goldwater and my son, Fred Eisenbud, whose suggestions have improved both factual accuracy and clarity of presentation. I am also indebted to Don Cioeta, an editor at the University of Washington Press, who fulfilled his role with skill, patience, and congeniality.

Chapel Hill   MERRIL EISENBUD
July 1989

AN ENVIRONMENTAL ODYSSEY

ONE

Random Walks and
Chance Encounters

A<small>N</small> INTRODUCTORY CHAPTER should set the stage for a story to be told, but my story did not exist in coherent form two years ago when I began to write it. It was scattered among fragments of memories, brief notes in many diaries and notebooks, an attaché case full of unsorted newspaper clippings, and in nearly two hundred journal articles and books I have published. Now that I have assembled the many pieces of my half-century odyssey, I am in a much better position to write this prologue to the story that lies ahead.

I started my professional career as an insurance company safety engineer. My first assignments were simple ones, such as inspections of small commercial establishments. Then I progressed to inspections of factories, to investigate not only accident hazards, but the dangers of occupational disease as well. I spent much of my time learning about the effects of dust on the health of workers and developing methods of dust control. In these tasks I found there were opportunities to do original work and I soon enjoyed the thrill a young man experiences from his first participation in a scientific meeting and having his research published in the journals of professional societies.

My career evolved through stages of increasing complexity, each of which provided preparation for the next. Yet, as in biological metamorphosis, each stage was indiscernibly related to the last. From insurance company inspector and safety engineer I progressed to industrial hygienist, research scientist, professor, environmental protection administrator, and advisor to more than twenty foreign governments, many agencies of

3

the U.S. government, the United Nations, and numerous industrial companies.

In all of this there was an element of chance in fortuitous encounters with both people and events. Professional acquaintances and colleagues, many of whom drift in and out of the narrative that follows, had major influences on my career, often unknowingly. However, their influence was secondary to that of three major historical developments in the mid-twentieth century. These were the sudden expansion of U.S. industrial capacity before and during the Second World War, the development of the military and civilian applications of nuclear energy, and the emergence of the grass-roots environmental movement. All three increased the importance of the work I was doing, while at the same time adding to its complexities and challenges. Were it not for this sequence of historical developments, I might have enjoyed a satisfying and productive professional career, but it would have been another life, in another field, and very likely would have lacked the many exciting changes that have resulted from the emergence of so many problems in environmental protection during my life.

Friends and colleagues have often asked me how the many parts of my professional life happened to fit together. I have never really known because, until now, I had never attempted to assemble the jigsaw puzzle. I have been too busy to look back, because it was more important that I look forward. Now that I have done so, it has proved to be a rewarding experience. I have always found my work enjoyable, but assembling these memoirs has made it doubly so, first in the doing and again in pleasant reminiscing.

Although I have been officially retired from university life for more than five years, I have been occupied by writing, speaking, and consulting. I now accept only those assignments that interest me, which is perhaps the greatest reward of active retirement. By being selective, it is possible to be surprisingly productive in a leisurely way when freed from the usual day-to-day chores and deadlines, as the following examples illustrate.

Since my "retirement" at the age of seventy in 1985, I have completed and published the third revision of my textbook *Environmental Radioactivity*.[1] The excellent reviews it has received have been gratifying.

In the fall of 1987 I appeared as an expert witness for the U.S. government in its defense against a class action suit brought by veterans who participated in atmospheric tests of nuclear weapons prior to the 1963 test ban treaty. About 220,000 veterans were involved in the tests, some of whom have developed cancer or have fathered defective children. How can one decide if the problems of the veterans who participated in the weapons tests are due to their radiation exposure or to the normal vicissitudes of life? Children with birth defects are born all too frequently to parents who have not been exposed to radiation, and about 20 percent of all Americans eventually die of cancer. These questions arise with increasing frequency in modern society, not only about radiation, but artificial chemicals as well. The expertise which the government wanted me to provide concerned the methods by which the dose received by the veterans was estimated. I chaired a committee of the National Academy of Sciences which investigated this question several years ago, and that was the reason for my involvement.[2]

In 1987 I was one of forty scholars from many countries who were invited to participate in a week-long Clark University centennial celebration symposium entitled "The Earth as Transformed by Human Action." It was an important conference, whose published proceedings[3] will, I hope, influence future policymakers as society enters the crucial twenty-first century, when the limited resources of the earth will be increasingly strained by a burgeoning population with an insatiable appetite for energy, air, water, food, and other natural resources.

At about the same time, my wife and I returned a few days early from a vacation in England because I had accepted a telephoned request from the governor of North Carolina to serve as chairman of a newly created Low-Level Radioactive Waste Management Authority. This authority is charged with the responsibility for developing a facility for storing the low-level radioactive wastes generated by the hospitals, nuclear power plants, and research laboratories in eight southeastern states that have formed a compact to provide for radioactive waste management on a regional basis.

Sandwiched between tasks such as these, I continue to be a member of a number of advisory committees, and have been invited to lecture in China, Japan, Austria, Brazil, Spain, and Switzerland. These and other activities which I will discuss in more detail in the proper sequence, have enriched my retirement years. I am grateful that good health, a long and happy married life, and my many associates in this country and abroad have made it possible for me to continue my life's work. My active retirement has meant that the hundreds of hours needed to prepare these memoirs have had to be found in competition with more immediate tasks. It would have been more logical to have postponed this book until more free time became available, but I was afraid that when that time came, I might no longer be equal to the task. It was either now or never!

I was born in 1915, in the middle-class apartment where my parents lived and my father practiced general medicine on the western edge of East Harlem, less than a block from Manhattan's Central Park. My mother, the former Leonora Koppeloff, was a woman of striking blond beauty whom my father met as the daughter of a patient. She was only twenty years old when I was born. I was the eldest of three children, with a brother and sister, Leon and Elsa. We grew up with lots of cousins who lived only blocks away and enjoyed the typical lives of the immigrant and first-generation New York families of that period. We lived in a mixed neighborhood and many of my playmates were the children of Irish and Italian immigrants, or of blacks who had recently moved north.

We often played in Central Park, not far from a lake on which we skated in the winter months and rowed in the summer. But the main playing fields for the children of the neighborhood were then on the streets, which as yet had few automobiles. In what should be better known as an example of human adaptability, generations of New Yorkers had evolved a variety of street games for children of both sexes and all ages. Any brick walls that abutted the sidewalks were used for handball. The steps that led to the houses were known by the Dutch word "stoops," and "stoop ball" was played by throwing a ball against them, with the object of catching it on the return bounce.

The streets were used by the older boys for roller skate hockey and various base-running rubber-ball games. The ball would be hit either bare-handed or with a broomstick. The ball would sometimes roll into a sewer, and its recovery called for mobilization of all the boys within shouting distance. Everyone knew the direction of sewage flow, so off would come the manhole covers at appropriate distances downstream. The big boys would lower the small ones into the sewer and the ball would be retrieved as it flowed by. A softball cost ten cents, which was reason enough to go to such trouble, but the main reason, I think, was the challenge presented.

I developed an early interest in science. I built my first radios when I was only eleven years old. I had already become expert at dissecting fish heads, chicken hearts, and various other kitchen offal. The mid-1920s was a period when there was a great deal of excavation for buildings and subways, and I became fascinated by the myriad crystalline forms in the newly blasted rock. I began to collect mineral specimens, a hobby that has continued to the present time. Now, more than 60 years later my collection includes nearly six hundred labelled specimens of cabinet quality from all over the world.

Those were the days before radio and television, but we never lacked for things to do. The nearby public library was well used. I enjoyed books about science and exploration. In 1925, when I was ten years old we moved across the Central Park to 87th Street, within walking distance of the American Museum of Natural History, where I began to spend much of my spare time. I found the late afternoon and evening lectures much more exciting than anything I was taught in school classes.

I was not a good student. My grades in mathematics and the sciences were very good, but it took me four years of high school to complete two years of French. I did not fail other subjects but I lacked the ability to concentrate on a subject that didn't interest me, so my grades in English and the social studies also suffered.

My entire formal education, from elementary school through college, took place in New York City. In June 1936, when I was three months past my twenty-first birthday, I was awarded a Bachelor of Science degree in electrical engineering from New York University. Like many college graduates of that period, I was pessimistic about the future. The country had not yet emerged from the Great Depression, jobs for young engineers were scarce, and war clouds were beginning to gather in Europe.

I was supposed to have prepared for medical school and had originally matriculated as a pre-med student but, for reasons that I have never fully understood, I changed to an engineering curriculum at the end of my freshman year. This I did without consulting my parents or anyone else. The fact that the engineering curriculum did not require foreign languages was surely an important factor. It was the kind of change that I should not have made without more advice than I received or asked for, but more than half a century later I can look back and say I have never regretted that decision.

As it turned out, I never earned a graduate degree. Although this was due initially to uncertainty about my career goals, later factors were also responsible, including the onset of the Second World War, the explosive growth of the field in which I found myself, and the excellent opportunities for on-the-job training under superb mentors who provided me with the knowledge and intellectual discipline I needed. On four occasions between 1936 and 1946, my plans to continue my formal education were frustrated by career opportunities. As a result, I may be the last—or am certainly among the last—of those who have been accepted into academic life and the world of science without having completed formal doctoral training.

I was self-conscious about this for many years since I was frequently addressed as "doctor" when I had no right to be. There were even times when out of necessity I was guilty of misrepresentation, as in the mid-1940s when I was lecturing at the Columbia University School of Public Health. The parking lot attendant would ask, "Are you a doctor?" and if I said no there would be no parking privileges for me! Later, in mid-career, I was fortunate to be awarded honorary doctorates that helped to remove whatever self-consciousness remained at that stage.

Although I believe I did not complete my graduate studies because circumstances made them either unnecessary or impractical, I may have been influenced subliminally by the experience of my father, who immigrated from Russia in the 1890s and became a respected physician despite a minimum of formal premedical education. He was the second oldest of three sons and, as in many Jewish families of that period, the men came to this country one at a time, usually in order of their age, so that they could earn money not only to bring their younger brothers and sisters to America, but also their cousins and other relatives. The first of the three brothers became a grocer and was sufficiently successful to help finance the ocean travel of the others. My father travelled alone to New York in steerage at the age of fourteen, having studied in a yeshiva to follow in the footsteps of his father and grandfather, both of whom were rabbis. To the dismay of his parents, he broke away from formal religion and found a job in an all-night drugstore when he joined those of his family who had preceded him to the lower East Side. To ease the crowding of the tenement in which the family lived, and for the convenience of the drugstore owner as well, he slept on a cot behind the counter.

In those days, the Board of Regents of the State of New York granted college equivalence diplomas based on examinations it administered. My father studied at work, passed the required Regents Examinations, and in that way became a licensed pharmacist. He continued his studies and eventually passed all the examinations required to qualify for entrance to medical school. In 1904, armed with his diplomas from the Board of Regents, he was admitted to Bellevue Medical College (now the New York University Medical Center) from which he graduated in the class of 1909, one year late because he had to take a year off for financial reasons. In the meantime, his younger brother had come to this country and eventually became a physician, as did the husbands of two sisters.

When I was growing up, I developed great admiration for the independence demonstrated by my father, and the ways in which he had successfully bypassed many formalities of the educational process. This may have been one reason why I did not give graduate education a high priority although circumstances definitely played the major role.

I spent the summer of 1936 beginning to get experience. My first job, for which I received twenty-five cents an hour, was on the 4:00 P.M. to midnight shift in a lower Manhattan loft where electrolytic condensers were manufactured. My title was "chief tester" (I was in fact the only one), and my assignment was to place a charge of several thousand volts on a rack of condensers which were then put in an oven as a sort of stress test. The condensers passed if they retained their charges after several hours. It was not much of a job, but at least it involved electricity! There being a depression, the working week consisted of only thirty hours, but there were no deductions in those days so my take home pay was exactly $7.50 per week.

I used the daytime hours to look for a better job and one month later was a foreman on the assembly line of a manufacturer of tabletop radios. My pay had doubled to fifty cents an hour and I was told I would work forty hours per week. As I later learned, the small factory was typical of many marginal businesses that were struggling to stay alive in those depression days. The twenty assembly line operators were taking home ten to fifteen dollars per week in exchange for which they worked in a badly illuminated and badly ventilated room with a filthy toilet and no hot water. As foreman, I stood between the submissive glad-to-be-working workers and an indifferent management who, to be fair, must have had many of their own problems in those depression days.

One afternoon I anticipated that the assembly line would be out of work early the next morning because a shipment of essential parts had not arrived. Since the hourly workers were not paid unless there was work to do, I suggested to the factory owner that it would be best if they did not report on the following day because we would certainly run out of the needed parts by midmorning. Nevertheless the owner decided to have the assembly line workers report, in the hope that the shipment would arrive overnight. This was unlikely, as I had been in touch with the supplier and was told that the shipment would be further delayed.

The following morning was a time that still comes to my mind when I experience a cold rain driven by high winds such as is frequently experienced in the New York area during the fall months. The assembly line workers arrived drenching wet from the subway kiosks and streetcars at eight o'clock and their work began promptly on schedule, only to stop ninety minutes later when we ran out of parts. The workers were paid for two hours work (about fifty cents) and sent home.

Nineteen thirty-six was a year of labor unrest; the electrical workers were being organized by a new union that was soon to merge with others to form the Congress of Industrial Organizations (CIO). Unknown to me, there had been some union activity in the shop and when I returned to work the next morning there was a picket line in front of the building. I don't know how things turned out, because I decided then and there to find another job. As one of the plant foremen I didn't want to join the workers on the picket line, but I was too much in sympathy with them to side with the plant management as I was expected to do.

By this time it was late September, and I decided to enter law school. This came about because an older cousin who had developed a successful patent law practice, and with whom I enjoyed a close relationship, thought I would be well suited to join his firm if I could combine law with my interest in science and engineering. Like many patent attorneys, this was what he had done. I was also impressed with the fact that Albert Einstein had begun his career in the Swiss patent office, not out of choice, but because, like myself, he couldn't find anything better to do at the time. Accordingly I matriculated into the evening division of the New York University School of Law. This made it possible to hedge my bets since I could continue to gain practical experience by day and could withdraw from law if something better turned up. That in fact is what happened, because of the first and most important of the many chance encounters that affected my career.

One day, as I was walking down Sixth Avenue (more recently Avenue of the Americas), I met an NYU civil engineering graduate two years my senior, with whom I had enjoyed frequent contact on the campus because we were both class presidents. I had lost track of him, but learned that he was working as a safety engineer in the New York office of the Liberty Mutual Insurance Company which was then, as now, one of the larger casualty insurers. He told me they were interviewing applicants for an entry-level position in their Philadelphia office. I had heard neither of the company nor of safety engineering but the opportunity sounded interesting, particularly when he told me that the starting salary was twenty-five dollars per week. We walked to his office and within a few minutes I was talking to his boss who called his counterpart in Philadelphia and made an appointment for me to be interviewed.

A few days later I was riding the Pennsylvania Railroad all the way to Philadelphia, ninety miles away, and at the expense of my prospective employer! That was a major adventure for someone who was born in East Harlem on the island of Manhattan and had never before seen any city but New York. My interview in Philadelphia went well. I was offered the job, and reported for work on the Monday following the Thanksgiving holiday of 1936.

That chance encounter resulted in the beginning of a long career. Because I happened to be on the right side of the street, walking in the right direction at exactly the right time, I met a friend and found a job. I thought it would provide me with a temporary salary and an opportunity to mark time while I decided what I really wanted to do. Instead, I was introduced to safety engineering, which soon led me to industrial hygiene, air pollution control, development of methods of protection against the effects of radiation, and the many challenging problems of environmental protection that emerged in the years ahead. Without realizing it at the time, I was being prepared to become a professional practitioner, a research scientist, a teacher, an administrator of technically complex programs, and a consultant and advisor to industry and governments in this country and abroad. In the process, I was to meet Irma, who has been my lifelong friend and wife of more than fifty years. We raised three sons who gave us three daughters-in-law and eight grandchildren. All because of that chance encounter in midtown Manhattan! I find it terrifying to think I would have missed all that has happened to me had I been on the other side of the street. Life is full of random walks and chance encounters.

TWO

Safety Engineer to
Industrial Hygienist

In 1936 THE Liberty Mutual Insurance Company was a relatively new organization, preparing to celebrate the twenty-fifth anniversary of its founding. It was best known to the public as a provider of automobile and home liability insurance, but it was also a major carrier of insurance coverage needed by industrial and commercial companies. Most of what I would be doing was related to their extensive program of workmen's compensation insurance.

I soon learned that insurance companies had played an important role in the field of public health for a century. Sir James Chadwick, who in 1842 wrote a classic report on the effects of slum living on the health of workers in England, was an insurance company actuary. The life insurance companies in the U.S. were well known for both the research they were doing on the causes of death and the educational programs they sponsored. The safety of electrical equipment had long been investigated by the Underwriters Laboratories, and insurance companies were also instrumental in the development of fire prevention codes. Insurance companies had thus long served an important social function in which government was not yet involved in a major way. The contributions of the companies to public well-being were not diminished by the likelihood that they were motivated by the increased profits that result when the number and severity of claims are reduced.

In keeping with this long tradition, Liberty Mutual maintained a large staff of technically trained personnel who inspected the facilities it insured and recommended ways to reduce the risk to the workers and members of the public. It was called the Engineering Department when I joined its staff in 1936, but the name was later changed to Loss Prevention.

I worked in a large office with perhaps a hundred desks belonging to members of the sales, underwriting, claims, and engineering departments. The one thing that I remember about my first few days was that the desk assigned to me was the only one not piled high with papers. I felt very unimportant as I sat down to an empty desk with nothing on it except a few reports given to me for orientation. However, that was not to last long.

Liberty Mutual proved to be a wonderful training ground for the career that was to evolve for more than fifty years. I spent eleven years with the company, of which the first four were concerned mainly with accident prevention and the remainder with industrial hygiene and the occupational diseases.

I was introduced into a fascinating new world during my first few months in Philadelphia. Until then my world had consisted mainly of my home and the schools I went to, all in New York City. Summer vacations were spent in "the mountains," no more than ninety miles from home. Except for a few trips through nearby factories and power plants as part of my undergraduate engineering training and the small electronic factories in which I had worked during the first few weeks after graduation, I knew nothing about the world of commerce and industry in which I would spend much of my life.

I worked with a friendly staff, most of whom had engineering degrees. For several weeks I accompanied them on their rounds of bakeries, laundries, shipyards, foundries, and even corner drugstores as they made their inspections. Of course, the country was still in the midst of a severe economic depression so that my first impression of the factories was that they were largely unoccupied. I vividly remember my first visit to a large machine shop which occupied perhaps a square block. The drill presses, lathes, milling machines, and other machine tools were almost entirely mothballed under heavy tarpaulins, with perhaps only one in ten machines in actual operation.

Insurance company personnel in those days were required to conform to a rigorous code of conduct. There was no smoking in the offices, and no tolerance towards deviations from what were considered the proper bounds of behavior. The most strict of the rules concerned dress. Hats were a must, and shoes should be shined but "not glaringly so," according to a memorandum distributed to all young engineers. That memo struck me as being so anachronistic even fifty years ago that I have kept it until the present. It provides an unusual insight into the decorum required in a large and successful company of that period.

The supervision was strict but compassionate. After a few days it became apparent that I would have to drive company cars to the many out-of-the-way places I would be visiting. But I had never learned to drive, having been brought up in Manhattan, where private cars were both a nuisance and unnecessary. This created quite a bit of amusement in the office, where the staff evidently had never known a college graduate who couldn't drive. Other members of the staff were instructed to give me lessons in the course of my daily orientation trips and it wasn't long before I was licensed to drive and could be on my own.

I was very fortunate that everything I was asked to do seemed interesting to me. This surprised me at first because the work at first involved no technical challenge. I used no mathematics beyond arithmetic, and had little need for my knowledge of electrical engineering. But I liked what I was doing because of the insight it gave me into the world of commerce and industry. The Liberty Mutual provided the insurance for industrial activities that ranged from abrasives manufacturing to zinc smelting. It insured the builders of the Pennsylvania Turnpike, then under construction, Gimbel's department store, and even the workers harvesting ice from lakes in the Pocono Mountains. My orientation program took me to large bakeries, shipyards, paper mills, steel mills, and machine shops, to mention a few examples.

Every day was exciting. Usually in the company of experienced safety engineers, I inspected large commercial bakeries and laundries, stevedores at work at the Delaware River piers, iron foundries, pharmaceutical manufacturers, furniture factories, and textile mills.

An assignment to the Gimbel's department store stands out in my memory because it was the first time I was on my own, only a few weeks after I reached Philadelphia. The Christmas rush was in its last days, and I was to spend full time at the store, inspecting it for possible causes of accidents among the shoppers. A surprising number of minor accidents were occurring each day, mainly because shoppers were tripping on imperfections in the flooring or slipping on stairways. Few of these incidents involved injuries, but all had serious potential. When I found something that required correction I wrote a recommendation on a pad of forms given to me, and then left the recommendations with the secretary of the store manager.

I had submitted perhaps one hundred recommendations to the store management by the end of the week, and I subsequently separated them into those that had and those that had not been properly dealt with. I found that the store management had complied with about half of the recommendations, but the remainder were uncorrected because the staff couldn't keep up with the number of recommendations I was making. Much to my surprise, I found that a considerable number of accidents, (fortunately all minor), were caused by the uncorrected conditions. This experience was gratifying and implied that at least as many accidents may have been prevented by the corrections made, since the deficiencies corrected were on average more serious than those that were not corrected. It was satisfying to find I could do some good in this way and I was left with a lifelong tendency to be on the lookout for accident hazards, not only in connection with my work, but in my personal life as well, often to the annoyance of my friends and family who have considered me somewhat of a nut on the subject of safety.

OCCUPATIONAL SAFETY IN THE 1930S

Not until the early part of the twentieth century did it become generally accepted that employers should bear the cost of injuries sustained by workers in the course of employment. Put another way, the costs of accidents in the workplace should be a proper charge against the cost of manufacturing a product or providing a service. Until then the financial burden of industrial accidents, including loss of wages, medical expenses, and permanent loss of earning ability was borne entirely by the employee and his family. This was the system that had evolved under common law, which held that the employer was not responsible for accidental injuries unless it could be demonstrated that he was personally negligent. Moreover, the employer could not be held responsible if it could be shown that there was negligence on the part of a fellow employee or the injured workman.

Under this system it was virtually impossible for an injured employee to be compensated for injuries that resulted from accidents at work. It was the rare case in which an employee could demonstrate negligence on the part of the employer without some contribution from a fellow employee or the injured workman himself. Moreover, the injured workman rarely had the knowledge or financial means to bring suit against his employer, and to do so would likely harm his future position in the company.

Thus, for about 150 years after the start of the industrial revolution, society allowed the human costs of industrial accidents to be borne by the injured worker. The personal hardships were borne by the workers and their families, but the financial costs were usually borne by the community to the extent that the employee and his family became public charges.

A system of no-fault workmen's compensation insurance was eventually developed that avoided the need to demonstrate negligence on the part of the employer, and thus eliminated costly legal proceedings in which the employee was inevitably at a disadvantage. The injured employee was compensated according to a published schedule of benefits without the need to establish fault on the part of anyone, in exchange for which the employee relinquished his rights to recover damages under common law. Since such damages were virtually unknown, workmen's compensation insurance provided by the employer resulted in a net gain for the worker.

The first workmen's compensation laws were passed in Europe in the early 1880s, but the United States lagged behind by about thirty-five years. A law was passed by the federal government in 1908, but applied only to government employees, leaving coverage for the private sector to the states. When several states passed the necessary legislation (e.g. Maryland 1902, Montana 1909), the laws were repeatedly declared unconstitutional. It was not until 1917 that the matter was finally settled by the Supreme Court, which ruled that the states had the power to establish a system of workmen's compensation. Individual states then began to adopt the workmen's compensation system, although the last state, Mississippi, did not do so until 1948.

In most states coverage was provided by private insurance carriers, among which Liberty Mutual was one of the largest. A better than average accident record resulted in lower premiums, which gave the employer an advantage over his competitors. In the more hazardous industries such as construction, the savings in the cost of workmen's compensation insurance could make the difference between profit and loss.

I entered the field of occupational health and safety during a time of transition. Apart from the fact that the last few states did not yet require workmen's compensation insurance, the laws were drafted in such a way that protection was provided for injuries by accidents in the workplace, but not for occupational disease. Accidents are the result of some un-planned occurrence, whereas occupational diseases usually result from the gradual and insidious action of a physical, chemical, or biological agent to which the employee is exposed in the course of his work. Exam-ples of these agents are lead, excessive noise, radioactive substances, and asbestos. Thus, a worker would receive medical care and compensation payments for being blinded by splashing drops of molten lead, but not for the neurological damage resulting from years of exposure to fumes emanated by the molten lead. The states were slow to modify their laws to include occupational disease, and the Pennsylvania law did not be-come effective until 1937, shortly after I arrived in Philadelphia. I soon found myself becoming more interested in occupational disease preven-tion than in safety engineering, which had to do largely with the preven-tion of mechanical accidents.

THE TRANSITION TO INDUSTRIAL HYGIENE

The profession that deals with the causes of occupational disease and the methods by which they can be prevented has been known as *industrial hygiene* and involves many disciplines. During the fifty years that I have been associated with the profession I have worked with people whose original training was in the engineering sciences, medicine, chemistry, bi-ology, meteorology, and even geology. The geologist, with whom I be-came a closely associated at Liberty Mutual, was a good example of how industrial hygiene drew specialists from many diverse fields of science, engineering, and medicine.

Charles R. Williams was a graduate student in the geology department at Harvard in the early 1930s, and did his doctoral research in mineral-ogy. His subspecialty was petrography, which is the art of microscopic identification of the individual mineral species by their optical properties. Williams became expert in the identification of the minerals contained in individual dust particles, which made him particularly useful to indus-trial hygienists who needed to identify the mineral species in samples of airborne dust, or dust in the lungs of industrial workers obtained at au-topsy. For this reason, Williams was invited to join the staff of the Har-vard School of Public Health, where the Department of Industrial Hy-giene was foremost among the research and training centers of that period. Williams also became a part-time member of the Liberty Mutual laboratory staff in Boston, where he later became head of the industrial hygiene group and an officer in the company. From his original interest in the optical characteristics of dust particles, Williams eventually became one of the world's leaders in industrial hygiene. Unfortunately he was a heavy smoker and succumbed to lung cancer at the height of his career.

The Department of Industrial Hygiene at Harvard was led by Philip Drinker, a member of a distinguished Philadelphia family who had graduated with a degree in chemical engineering and was invited to Harvard by his brother Cecil, a well-known physiologist, to design and build a device to assist people suffering from respiratory paralysis. This was particularly important in cases of infantile paralysis (poliomyelitis), which was a dreaded disease until it was eradicated by the vaccines developed by Salk and Sabin. The device was successful and was known for many years as both the Drinker Respirator and the Iron Lung. Philip Drinker remained at Harvard for many years and developed a major center for training occupational health specialists. Liberty Mutual maintained a close relationship with Drinker's program.

Although I was stationed in Philadelphia for my first few years with the company, its headquarters was in downtown Boston, where a small laboratory was maintained on the ninth floor of their building, at 175 Berkley St. The laboratory was directed by Stuart Gurney, a chemist who provided consulting services to safety engineers throughout the country. Much of the work of the laboratory, which in those days employed only four or five technicians, was concerned with identification of toxic substances in materials used in industry. There were no labelling laws in those days, and solvents or other materials used in industrial processes were frequently known by a trade name or a designation such as WX-64. Since the composition of the materials was proprietary information that often was not revealed even to the insurance company, it was necessary to obtain samples for analysis by the laboratory.

It was frequently necessary to collect and analyze samples of gases or dusts present in workroom air to determine the amounts of hazardous substances to which the employees were exposed. When this was required, Gurney or one of his assistants visited us in the field and joined us in the collection of the samples for analysis. These field trips were my first hands-on experience with industrial hygiene.

I should note that the literature of that period in the field of industrial hygiene was not extensive. There were five or six textbooks with titles such as *Industrial Dust* and *Industrial Toxicology*. The most important source of information was the *Journal of Industrial Hygiene and Toxicology*, which began publication in 1916. The journal carried not only articles, but also abstracts of other relevant publications in medicine, engineering, and chemistry. Other important sources of information were the bulletins and reports of the U.S. Bureau of Mines and Public Health Service. All of this literature could be accommodated on about two five-foot shelves; almost all of it was located conveniently only a few yards from my desk in the office of my supervisor.

I found it an easy literature to digest and within six months I had read it all and was eager for more knowledge about occupational diseases and how they could be monitored and controlled. In this I probably benefited from the fact that I had grown up in a home in which my father practiced medicine. I had been a science buff since I was ten years old, and since I had assumed with the rest of the family that I was preparing for a medical career, I had spent many hours listening to my father discuss medical matters with other physicians. I had also avidly read his books and journals until I left home at the age of twenty-one. Although there was much I could not possibly understand, I at least had understood the basic language of medicine, which was an important advantage when I became involved with industrial hygiene.

By 1939, three years after I started to work in Philadelphia, I had made up my mind that I wanted a career in industrial hygiene. I knew very little about the situation in other states, but I could see that much needed to be done to reduce occupational disease in Pennsylvania, and that Liberty Mutual was in a good position to provide leadership. None of the organizations that we now rely on to improve conditions in the workplace existed. The occupational disease amendments to the Pennsylvania workmen's compensation law had just become effective, and the state had not yet developed the kind of industrial health surveillance functions that now exist. The only involvement of the federal government was through the Bureau of State Services of the Public Health Service, whose role was advisory, and that only when its assistance was requested by a state. The insurance companies not only had unrestricted access to the factories they insured, but they carried a big stick in the form of their ability to cancel insurance policies or increase the premiums if conditions were found to be substandard.

There were several early experiences that made deep impressions on me. The stories are interesting in themselves and serve to illustrate the state of industrial health at the time, as well as the broad range of activities with which I became involved.

Liberty Mutual insured two tanneries, both of which processed goatskins imported from Asia. It was well known that handling Asian skins subjected the workers to the risk of developing anthrax. An anthrax infection begins as a small painless pimple which, if not promptly treated, can spread to the entire body with a high risk of death. Those were the days before development of the antibiotics and sulfonamides, which have proven to be highly effective against the anthrax bacillus. The arsenicals were being used, with somewhat inadequate effectiveness, particularly if not administered in the early stages of the infection.

The Philadelphia area was an important leather producing center, and the two companies were on the Camden side of the Delaware River. Both received goatskins from identical Asian sources and processed the skins in the same way. However, there was an important difference in that plant A had experienced no serious cases of anthrax in many years, but plant B had experienced cases every year or two, some of which had terminated fatally. I was asked to study the two plants in the hope I could identify the reason for this difference.

After several days of study I could find only one relevant difference between the two plants. Plant A, which had the better record, employed a nurse who had excellent rapport with both the employees and the physicians retained by the company for consultation. She had taught the workers to examine themselves regularly, and to come to her for examination whenever a new pimple appeared. This involved quite a workload, because there were a few hundred workers at risk and pimples unrelated to anthrax appeared among them not infrequently. She examined the pimples and referred the employee to the consulting physician if she thought there was any possibility that the pimple might be the start of an anthrax infection. Plant B also employed a nurse but she had not developed a similar educational and screening program. The exemplary record of plant A seemed to have resulted from the initiative of the industrial nurse who understood the importance of early screening and referral to a physician for treatment.

Several months later I was invited by the Philadelphia Academy of Medicine to participate in a panel on industrial anthrax. This was my first invitation to contribute to a scientific program, and I was delighted to be able to report on the results of my brief study.

A second experience had a much greater impact on me. I had inspected a chocolate factory in a routine way, looking for unguarded machinery and deficiencies in the plant housekeeping. One of the hazards of the manufacturing process was that dry chocolate dust could accumulate on rafters and in ventilation ducts, particularly near the machines that grind the beans. If this dust became suspended by sudden vibration of the building, the airborne chocolate dust might ignite, resulting in a destructive dust explosion.

This particular plant had had a good safety record and previous inspectors had given it a clean bill of health. So did I. However, much to my surprise, two cases of silicosis were reported about one year following my inspection. This seemed impossible, because silicosis develops from inhaling particles of silicon dioxide (silica) and I was unaware that any of the workers were so exposed.

But I was wrong. Early in the chocolate making process, the chocolate beans were ground between granite milling stones, and from time to time the stones required dressing. This was done in a small room I had not seen, using pneumatic chisels that generated considerable amounts of dust. Since granite contains silica in a form that can produce silicosis, and since the workroom was not properly ventilated, a hazardous condition existed. I had known that there were granite milling stones being used in the factory but it never occurred to me that they required dressing. That experience taught me the need for thoroughness in the inspections I was conducting. Unhappily, we all seem to learn from bitter experience.

There was still another memorable early experience. A workmen's compensation claim for silicosis was filed by a shipyard worker who had a history of sandblasting. This is usually done with beach sand that is almost pure silica, and it appeared that the claim might be on solid ground, especially based on the seemingly conclusive reports from his physician, who was well known for his interest in occupational lung diseases. However, when I investigated the matter I found that aluminum oxide rather than sand was used for the abrasive. Aluminum oxide does not cause silicosis. I could find no evidence that the worker had ever been exposed to silica, either on that particular job or any other. On the basis of my findings, and the report I submitted, the claim for silicosis was denied.

A few days later I was surprised to receive a telephone call from the employee's radiologist, Eugene Pendergrass, the Chairman of the Department of Radiology at the University of Pennsylvania Medical School and a respected silicosis authority. He told me that my report was in error and that there must have been a history of exposure to silica, because the worker definitely had silicosis. He asked that we discuss the matter further and suggested that we meet at the University of Pennsylvania Hospital, where the employee was confined, in the hope we could obtain additional information from him. However, the man was already close to death, and we refrained from talking to him, but his wife was in the room and we reviewed her husband's working history with her. She could not provide information other than what I had already reported, so it was only a brief conversation. However, the interview stands out in my memory because of her bitterness towards me. She regarded me as a dishonest insurance inspector who was about to deprive her of what was surely a much needed workmen's compensation award.

I must confess that at that point I could not help but realize how easy it would have been to change my report. I had only to report a history of "probable" exposure to silica, and the award would have been hers upon the death of her husband. This would have been easy for me to have done, especially in view of the findings from Dr. Pendergrass. However, I had investigated the occupational history of the worker thoroughly, and he had not been exposed to silica. Since the worker had only a few days to live, it was clear that the decision could soon be based on post mortem findings if permission for an autopsy could be obtained. Dr. Pendergrass explained this to the wife, and she readily consented.

A week or two later I received a call from Pendergrass in which he informed me that his patient had died and that the cause of death was not silicosis, but a type of lung cancer that could be confused with silicosis in X-ray examination. I received the news with mixed feelings. I was glad my opinion was validated and that the death was not due to the negligence of the company insured by my company. However, I was sorry that the worker's widow would not receive badly needed financial assistance. I would probably have forgotten the incident long ago were it not for the fact that Dr. Pendergrass, forty years my senior and one of the country's leading authorities on occupational diseases of the lung, had the kindness to call me to report that his diagnosis was wrong and that I was correct.

Because of my growing interest in the field, I was anxious to obtain formal training in industrial hygiene. Theodore Hatch, who had been associated with Philip Drinker for many years at the Harvard School of Public Health and was coauthor with Drinker of *Industrial Dust*, came to the University of Pennsylvania in 1939 to develop a teaching and research program in industrial hygiene. I visited him to inquire about my entering his graduate program. He gave me encouragement, and with the concurrence of my supervisor I wrote to a vice-president at our home office in Boston to ask if the company would give me leave to enter graduate school, with the understanding that if they continued to support me financially, I would return to employment with them on completion of my degree.

By this time it was early 1940, there was a major war in Europe, and production of the goods of war in this country was increasing. Our military was also slowly mobilizing and Hatch, who held a reserve commission, was called into the army, which ended my plans to enter the University of Pennsylvania. For a while I explored the possibility of matriculating into the Harvard program, but the company decided that things were beginning to move too fast. While the idea of going to graduate school was a good one in principle, there was no time for it. Instead I was to be one of four safety engineers selected to participate in a six-week intensive course being prepared at Harvard. It turned out that this was a sensible compromise. Although six weeks was hardly a substitute for a formal graduate program, we were all motivated and receptive students and probably learned as much in those hectic six weeks as would have been possible in a year or more under normal circumstances. They kept us so busy that we were not allowed to return home for Thanksgiving, although they did allow Irma, my recent bride, to visit Boston for a one-day holiday reunion.

There was a meeting of the New England Chapter of the newly formed American Industrial Hygiene Association in Portsmouth, New Hampshire one weekend while our course was in progress, and students and faculty drove together to attend. It was a small meeting, because there were so few people interested in industrial health, but I was impressed with their friendliness. This was my first opportunity to become acquainted with many of the leading scientists, engineers, and physicians in the field. I was pleased to find that Liberty Mutual was highly regarded, and that as a member of its staff I was welcome among these experts. I was to find later that my association with Liberty Mutual made me welcome wherever I went during my remaining years with the company.

I returned to Philadelphia with the title of Industrial Hygienist and the assignment to serve as a consultant to the safety engineers wherever I was needed. One of the four men who received the specialized training remained in the Boston headquarters, and two others were soon drafted into the military service. Charles Williams had become our supervisor, and only he and I were available to fulfill the industrial hygiene needs of the many field offices. It was the beginning of a four-year period in which we would be called upon repeatedly to investigate the health problems arising in the expanding war industries.

THREE

Industrial Hygiene in the War Industries

On DECEMBER 7, 1941, the Japanese attacked the U.S. fleet at Pearl Harbor and the U.S. once again became a participant in a global war. At age twenty-six, and in excellent health, married but childless, I applied for a commission in the navy but was told within a few days that I could not enter service because my work with Liberty Mutual was considered important to the war effort. This came as a big surprise to me and was at first disappointing, but the decision proved to be immensely beneficial to my career while at the same time giving me a genuine opportunity to contribute to the civilian war production program.

Those responsible for war production were apparently more aware than I had realized of the health problems that had developed among the factory workers during the First World War. There had been many tragedies. In the manufacture of the explosive TNT, more than seven hundred employees had died of liver disease caused by their exposure to the toxic vapors produced during production. Many young women who had painted timepieces and compasses with luminous paint that contained radium began to die in later years from bone cancer caused by radium that had been absorbed into their bodies and deposited in their skeletons. Men who had worked under water in pressurized caissons suffered from "the bends," a painful and disabling disease caused by the formation of nitrogen bubbles in the body when the men returned to the surface. The tragedies of the workplace were insignificant in numbers compared to those in the trenches, but they were preventable and contributed to production inefficiencies. It was recognized that there was a close correlation between a factory's safety record and its production efficiency.

Liberty Mutual provided workmen's compensation insurance for some of the largest companies involved in the war effort. The extent of its involvement was illustrated by the fact that two Liberty Mutual safety engineers, who had been temporarily assigned to inspect military construction on Wake Island in the central Pacific, were captured when the island was surrendered to the Japanese shortly after war broke out. The two engineers spent four years of the war in a prison camp in the Philippines.

After nearly five years in Philadelphia, it was decided that I could operate more efficiently out of the New York office, to which Irma and I returned shortly after the outbreak of war. In addition to the fact that we would be nearer to our families and friends, I was pleased at the prospect of being in New York because it was a center of industrial hygiene activity. The state of New York maintained a well-known laboratory in Manhattan, as did the Metropolitan Life Insurance Company.

I had much to learn about all kinds of new industrial processes, some of which were still in the development stage. When one thinks about war industry, it is natural to visualize aircraft production, shipbuilding, tank assembly lines, and gun factories. It was these and much more. The supplies of rubber from Southeast Asia had been interrupted by events in the Pacific, so a new synthetic rubber industry was needed. Penicillin, the first of the antibiotics to be developed, was needed in huge quantities. New aluminum extraction plants were required and would be built near the sources of hydroelectric energy. Many of the employees in these new industries would be exposed to new and poorly understood dangers of occupational disease. It was the job of the industrial hygienist to identify the hazards and make recommendations for their control before any damage was done.

During the war years I depended heavily on information from more experienced industrial hygienists. In addition to those located in New York, I had ready access to the pool of expertise at Harvard, where Liberty Mutual enjoyed excellent relationships. We also depended heavily on the staffs of many of the companies we insured. If I wanted to measure the concentration of a new pharmaceutical product in the workroom air, I knew how to collect the sample, but who was in a better position to perform the difficult analysis than the research chemists working for the pharmaceutical company? More routine types of analysis could be done in our Boston laboratory, but it was impossible to develop all of the sophisticated types of analyses that were required. Liberty Mutual and the insured companies had a common interest in wanting to avoid diseases of the workplace.

WORKING IN THE FIELD

The territory I covered included most of the United States east of the Mississippi River. The facilities I visited were varied and the time I spent in them could be as short as a few hours or as long as two weeks. Frequently I was required to examine plans for a new process, identify toxic materials that required special care, and make specific recommendations for their control.

There was an extraordinary amount of learning to do. My reading included textbooks on chemical and mechanical engineering, the *Journal of Industrial Hygiene and Toxicology,* and any other useful sources of information I could locate. I was amazed at the extent to which I could obtain help from other people, provided I made a conscientious effort to help myself before coming to them. They were willing to take the time to help me with explanations and literature. I quickly learned that in my work I had to know how to critically define the limits of my own technical knowledge, never hesitating to admit when that limit had been reached, and always seeking help when I needed it.

Apart from the purely technical aspects of my work, I found mastering the languages of their trades helpful in dealing with the factory foremen and supervisors. A kiln in the ceramics industry was everywhere referred to as a "kill." Only a novice would pronounce the final consonant. I would be thought more sophisticated about shop matters if I referred to a certain machine used for filtering slurries as the "Oliver," (if that indeed was its manufacturer), and if I knew that the muck they collected on the cloth filtering material was the "raffinate." The entrance to a mine was the "adit," the tunnels were the "drifts," and the working face of the mine was the "stope."

Much of my time in the field was spent collecting samples of air for analysis, since the risks of many occupational diseases depended on how much of a particular substance was inhaled by the worker. The equipment available for air analysis has improved enormously since those days, and there are now compact instruments that are light and provide on-the-spot measurements of many toxic substances. This was not possible in the 1940s, when the standard procedure was to bubble air through water in a flask known as the Greenburg-Smith impinger, which had been invented at the Bureau of Mines in the 1920s. Most of the samples were collected in the field by the impinger or some similar device and shipped to the Boston laboratory for analysis.

One major exception was the measurement of the concentration of insoluble airborne dusts such as the mineral dusts encountered in many industries. For this purpose, a pump that weighed about fifteen pounds and a box of impinger flasks were taken into the field along with another box of accessories. In the course of a day, perhaps a dozen samples were collected, which then required "counting." When away from New York, this was best done in my hotel room to minimize the chances of sample contamination. The counting procedure involved pipetting an aliquot of the suspension onto a slide identical with those used for blood counts. I thus spent my night hours peering through a microscope (which I also carried with me), to count the number of dust particles contained in the sample. Since the volume of air from which the sample was collected was known, the concentration of airborne dust, expressed as millions of particles per cubic foot of air, could be calculated easily.

Each hazardous substance was assigned a "maximum permissible concentration" (MPC), and whether the workers were at risk was judged by comparing the measured values with the recommended permissible limit. What we then called the MPC is now referred to as the Threshold Limit Value (TLV), and the values are much lower for most substances.

It was necessary for me to take other equipment into the field as well. For measuring the velocity of air movement, which was frequently done to judge the efficiency of ventilation systems, we used an instrument known as a "velometer." For measurement of the presence of explosive gases, a flammable-gas detector was used. It was one of the few field instruments that gave immediate results. Other instruments I used were the mercury vapor detector, which operated on the principle of ultra-violet absorption, and the carbon monoxide detector, which, like the mercury vapor detector, was a direct reading instrument contained in a black box slung over the shoulder. Finally, it was frequently necessary to collect an airborne dust sample in a dry condition, in which case we used an electrostatic precipitator, which deposited the dust on the inside surface of an aluminum cylinder.

This equipment was bulky and heavy. Had it been possible to plan sufficiently in advance, the equipment could have fitted easily into a station wagon, but during the war years we spent much of the time dealing with emergencies, and since the distances were often long, it was necessary to make the trip by overnight train. Often we didn't go from place to place in any orderly fashion. I might be in St. Louis one day, then travel to Philadelphia, only to find after one day that I was needed in Chicago! I spent many nights in Pullman cars crisscrossing the country. At each stop I was faced with the need to move my equipment from the railroad car to the hotel with a near absence of porters, and a wartime scarcity of taxicabs.

INDUSTRIAL HYGIENE PROBLEMS IN THE DEFENSE PLANTS

Some of the industrial hygiene problems that developed during the war were dramatic and not all of them were dealt with in a manner that would be thought adequate by modern standards. Many of the episodes went unreported in the scientific literature, not because there was a reason for secrecy, but because of the hectic pace of the wartime years. Also, for lack of space, the journals that then existed were unable to accommodate scientific reports of everything worthy of publication.

Thus, so far as I know, there is no comprehensive report of the wartime experience with Halowax, an electrical insulating material that was widely used. Halowax was a mixture of chlorinated diphenyls and naphthalenes that was not only an excellent insulating material but was also fireproof. It was specified for use in all cables installed in Navy ships. Other Halowax formulations were used in electrical condensers that found their way into military electronic equipment of all kinds. Unfortunately, Halowax was highly toxic to those who worked with it. In two plants with which I was familiar, nearly all of the workers developed extensive skin eruptions known as chloracne. More serious was the toxic effect of Halowax on the liver, and a dozen workers in the two plants died of liver disease as a result of their exposure.

Today we would not hesitate to shut down such a plant and possibly ban the product. An analogous and more recently publicized family of compounds, the polychlorinated biphenyls (PCBs), which are closely related to the ingredients of Halowax, also cause chloracne and can damage liver function. (PCBs may also be human carcinogens, although this is debatable.) PCB manufacture and use has been banned in the United States since about 1970. However, in 1944 one had to be circumspect about a matter like this. Most of the Halowax was being produced to insulate the electrical cables being installed in ships. In a vessel aflame in battle, it was essential to maintain shipboard communications.

We never really understood what needed to be done to control the problem in the cable plants. We did the best we could by improving the plant ventilation systems, and by adopting procedures that minimized contact with the Halowax. The difficulties were reduced but never really solved until nontoxic substitutes were developed late in the war.

Shipbuilding was one of the most important war industries and unfortunately was one that left a legacy of occupational disease that affected the shipyard workers in subsequent decades. Asbestos, a mineral fiber that had been used for thermal insulation for centuries, was widely used aboard ships to insulate steam pipes, boilers, and other sources of heat. It had been known since the turn of the century that inhalation of asbestos fibers could cause a serious lung disease known as asbestosis, and maximum permissible concentrations of asbestos dust had for some years been in use by industrial hygienists. What we did not realize at the time was that the asbestos fibers were also capable of causing two kinds of malignant disease, cancer of the lung, and mesothelioma, a cancer of the pleural and abdominal linings. We have also since learned that the maximum permissible concentration used for asbestos during the war years, which was designed to prevent asbestosis, was too high and required substantial downward revision.

The asbestos story has been complicated by the fact that many of the cases developed decades after exposure and there frequently had been more than one employer involved. Employers and insurance companies quarreled about which of them was responsible. Another problem is that the system of workmen's compensation insurance that had served, however inadequately, as a no-fault avenue of redress that avoided costly litigation, has given way to new theories of law in which the company that supplies the toxic product can be held responsible for injuries to employees of companies that use the product. A further complication is that asbestos was used in wartime shipbuilding because it was required by the U.S. Government, which a citizen cannot sue without its consent. Up to the present time, the government has not accepted responsibility for the damage done by the asbestos it required. In the years since World War II, thousands of claims for asbestosis and cancer have been made against the asbestos manufacturers, the largest of which has declared bankruptcy.

Another problem encountered in the shipyards was far less serious in terms of the ultimate effects on the workers, but was far more disruptive to production schedules. This was "metal fume fever," which occurred among workers who welded galvanized steel, which is coated with metallic zinc. Unlike asbestos dust, which produced its effects after many years, the zinc fumes caused a severe but short-lived fever a few hours after exposure. The employee was left in a weakened condition following such an episode and, although there were no known aftereffects, the cases were sufficiently frequent that there was a need to develop preventive measures.

For most work this consisted of having flexible exhaust hoses that the welder positioned at his work. However, it was not practical to use the four-inch-diameter flexible steel hose under all conditions. A notable exception was in the inner bottoms of merchant ships, the hulls of which were honeycombed with small steel compartments that were of welded construction. Because the system of exhaust ventilation through the flexible steel hoses was not practical there, I designed a welder's helmet in which the face was a perforated plenum into which compressed air could be pumped, thus bathing the welder's face with fresh air. This was a useful development which my employer encouraged me to patent.[1] I was also allowed to retain the royalties from sale of the helmet, which was a very generous decision on the part of the Liberty Mutual. I took the helmet to Harvard and tested the device with Leslie Silverman, who was then a young instructor in Drinker's department. We received an award from the American Welding Society for the paper we prepared based on these tests.[2]

The sudden requirement for tens of thousands of aircraft required new plants to produce aluminum. The aluminum production process resulted in the release of gaseous wastes that contained fluorides, as a result of which pasture land near some of the plants became so contaminated that grazing cattle developed fluorosis, a crippling bone disease. This is but one example of many episodes that would be given widespread attention by today's more environmentally concerned media, but went unnoticed in the midst of the tribulations of the wartime period.

We never knew where the next surprise would come from. For example, who would have thought that cashew nuts could present a major health problem? We all know these nuts as benign and tasty tidbits, but we enjoy them after removal of the outer husks with which they are harvested overseas. When we had reports of an epidemic of disabling dermatitis among stevedores unloading a ship from South America, I was sent to investigate and found that they were handling a cargo of cashew nuts as they had been gathered in the tropical forests. The nuts we eat grow in large numbers within husks, between the inner and outer shells of which is a spongy lining containing an oil used for a varnish needed in the manufacture of certain electrical components. Unfortunately for the stevedores, the oil is very irritating and produces a rash similar to that caused by contact with poison ivy. By the time we received the reports of the dermatitis outbreak, the cargo had been unloaded, but I arranged for the consignee to send a sample of the nuts to our laboratory in Boston in the hope we could identify the source of the problem.

I would have been wiser to have discussed the matter first with the varnish factory that was receiving the nuts, since they knew of the problems with cashew nut husks and were aware of the special handling precautions required. Instead, the nuts were sent by the shipping company to Boston without any warning and caused a mini-epidemic among our small staff of laboratory technicians. This was one of my more inglorious contributions to the war effort.

One of the companies we insured had the patent on the mercury battery, which is now well known as the dime-like cell used in watches, hearing aids, cameras, and hand calculators, which need a power source more compact than the batteries then used in flashlights or portable radios. Production of mercury batteries began in 1943, and they proved so useful to the war effort that several more plants were constructed for their manufacture. These were producing 1.5 million batteries per day by 1945.

Mercury is the only metal that is liquid at room temperatures. As an object of fascination since ancient times because of its beautiful color, the popular name "quicksilver" is reported to have been given to mercury by Aristotle because it was believed to be "living silver" in the purest of forms. Because of its high vapor pressure, mercury is released to the workroom atmosphere in the form of vapor when the metal is exposed to the atmosphere.

Although mercury is mined as the sulfide mineral cinnabar, which is relatively non-toxic, tiny globules of the metallic element are present in the ore in sufficient amounts to contaminate the air of mercury mines. This has been known for centuries in Almaden, Spain, where mercury has been mined for 2,400 years. The toxic effects among the Spanish miners have been prevented by limiting the time a worker spent underground to 32 hours a month.

High levels of mercury exposure can cause kidney disease, but neurological damage can be caused by less severe exposure. Classic symptoms

of mercurialism were endemic in the felt hat industry for many years because mercury nitrate was used in the process. Its use was eliminated after a U.S. Public Health Service study in 1941 reported extensive disease
among the workers. (It is widely said, perhaps incorrectly, that Lewis
Carroll's description of the Mad Hatter referred to the characteristic irri-
tability and tremors seen in hatters as a result of their exposure to mercury.)

We recognized that these batteries required large quantities of mercury
and because of the bad previous experience with the metal, we gave a
high priority to advising the policyholder about techniques by which the
batteries could be manufactured safely. As a result, tons of mercury were
used without health effects, using methods that our staff developed in cooperation with the manufacturer.[3] This was an excellent example of the
principle that a hazardous substance can be used safely in industry if appropriate precautions are taken.

There were other success stories. During the First World War radium
had been used as the principal ingredient of luminous paints applied to
watch dials, compasses, and other instruments used by the soldiers and
seamen. There was little attention paid to industrial hygiene at the time
and primitive work habits were permitted. The dial painters were mostly
teenage girls paid by the number of pieces completed. The young ladies
soon found that they could worker faster if they pointed the brushes with
their lips, which resulted in the ingestion of small amounts of the radium.
This would have been a poor hygienic practice for individuals working
with even the most innocuous of industrial chemicals but radium proved
to be a very hazardous material because of its radioactive properties.
Many years after their wartime work some of the women began to develop bone cancer because radium is chemically similar to calcium and
when ingested deposits in the skeleton, from which it is eliminated very
slowly.

The early cases involved disease of the jawbone and were identified by
a dentist in northern New Jersey, where the largest of the luminous dial
factories was located. Some of the bone changes developed into cancer.
The dentist was at first unaware that the paints contained radium, and
certainly had no reason to associate ingested radioactivity with cancer.
He at first confused the luminescence of the products with phosphores-
cence, and assumed the women were suffering from "phossy jaw" which
was a disfiguring disease seen among workers in match factories who had
been exposed to a particularly toxic form of elemental phosphorous.

That radium rather than phosphorous was responsible was first recognized by Harrison Martland, a pathologist who undertook the first of several important studies of the luminous dial painters, starting in 1925.

The fact that radium could cause cancer created a sensation, and for a while there were many who doubted that this was possible. Among those was Madame Marie Curie, who had discovered radium about twenty years previously, and now asserted that radium cured rather than caused cancer. Martland once showed me a letter from Curie in which she called him a charlatan for writing that radium was responsible for the cases of bone cancer. I made an effort to find the letter when Martland died but was unsuccessful. Perhaps it will turn up some day.

In all, only a few hundred grams of radium had been extracted from the earth's crust and this small amount was responsible for the deaths of at least a hundred people. Not all of these were radium dial painters. In the early part of this century, radium was used as a nostrum for many diseases, including arthritis, syphilis, and even insanity. Some of the patients to whom the radium was administered also developed bone cancer.

The radium tragedy resulted from ignorance about the effects of overexposure to radioactive materials. Fortunately, the misuses of radium were so well studied after the effects became known, that relatively large quantities were used without known injuries during the Second World War. Strict rules of personal hygiene were observed, and the work was performed in well-ventilated rooms. Much of the credit for the safe handling of radium goes to Robley D. Evans, an MIT physicist who became interested in the subject of radium poisoning in the 1930s. Based on his work, standards for the safe industrial use of radium were adopted in 1940 and exist to the present time. One of the many interesting properties of radium is that it decays to radon, an inert gas. When radium deposits in the body, the radon produced can be measured in expired breath, and this was the method we used to determine the extent to which the radium workers had accumulated skeletal deposits of radium. Samples of breath were collected in one liter flasks, which were then sent to a laboratory for radon analysis. This method was used until about 1960, when whole-body counters were developed that used gamma radiation as the basis for estimating the body burden of radium.

The experience with the medical and industrial uses of X-rays was also tragic. Many of the radiologists and physicists who worked with these rays soon after they were discovered in 1895 developed cancer of the hands and other parts of the body. By 1928 the dangers from the use of

X-rays and radium were sufficiently well recognized that an International Commission on Radiological Protection was established and continues, to this day, to recommend the safety standards used world-wide. The National Council on Radiation Protection and Measurements was established in the U.S. in 1929. These two organizations have provided the world with the guidance that has made it possible to benefit from the medical, research, and industrial uses of X-rays, radium, and other sources of radiation while maintaining an excellent record of safety.

Aircraft production took the largest portion of my time during the war years. The industry exploded into an enormous complex of mostly new manufacturing plants that incorporated the best architecture, were well illuminated, and had fine cafeterias (which was a novelty in those days). But it was a complex industry with many potential health hazards.

The planes manufactured ranged from the light aircraft used for courier and for training service, to fighter planes and the B-29s that did most of the strategic bombing. Curtis-Wright and Pratt and Whitney built most of the reciprocating (piston) engines that were needed. (Jet engines were still experimental.) Both companies were insured by Liberty Mutual, as were some of the larger airframe manufacturers.

One of the problems that concerned us in the engine plants was the noise produced at the test stands. Noise was of course a problem in the shipyards, the airframe construction plants, and elsewhere in industry, but at the start of the war, the noise created during the testing of the reciprocating engines was of particular concern to us because workers in the older plants had suffered loss of hearing. However, it wasn't long before acoustically insulated test cells were designed that protected the workers.

Plastic ear plugs molded to individual ears were a practical solution to the noise of rivetting. When this technique was still in an experimental phase I fitted a great many employees for their earplugs and became very unpopular because the plastic hardened around the hairs of the outer ear, which made it painful to remove the plugs the first time they were used. In addition to being noisy, the rivetting tools frequently caused a painful and temporarily disabling inflammation of the tendons.

The instrument panels used during engine testing were equipped with many manometers that contained mercury. These created a problem because the metal occasionally spilled and contaminated the cracks and crevices of the floors with tiny globules of mercury that were a continuing source of exposure to mercury vapor. The globules were difficult to recover, but we solved the problem by treating the floors to convert the globules of metallic mercury to a nonvolatile sulfide. This effectively eliminated the problems caused by volatilization of the elemental mercury. Other health problems with which we were concerned in the engine plants were caused by electroplating, exposure to fluorides used in the magnesium foundries, and X-Rays used in inspection of the castings.[4]

There were many other potentially hazardous procedures that required investigation in the aircraft (airframe) manufacturing industry, including plating, degreasing, and paint spraying. The aircraft were camouflaged early in the war, which required that a zinc chromate priming coat first be applied and then the multicolored camouflage coat. Zinc chromate is an irritating compound, and we have since learned that some of the chromate compounds are carcinogenic, although there is no evidence of an excess of cancers among the sprayers. The solvents and thinners used in the paints also required precautions, since some were toxic. Many of the smaller subassemblies could be painted in ventilated booths, but this was not possible for the entire airframe assembly. Instead, the workers, mainly women, used respirators connected by a hose to a source of compressed air. This was really a clumsy arrangement, because the women needed to climb all over the planes. Fortunately it was soon discovered that the weight of the paint slowed the aircraft enough to offset the advantage of the camouflage, so the painting was discontinued.

I had one unusual experience that did not involve occupational health. The company insured the manufacturer of a precision instrument used to assist the accurate placement of bombs dropped from B-29s. The instrument included a small ball bearing no more than a quarter inch in diameter. It was learned from field reports that dust in the bearings was impairing the operation of the bombsight. After finding that the dust particles were probably entering the bearing during its manufacture, extreme precautions were taken to reduce the airborne dust where the bearings were assembled. A special "clean room," the first of its kind, was constructed in which the air supply was efficiently filtered. The female workers were required to wear dust-free plastic garments and hair covers. Yet the problem persisted.

Since I was developing a reputation as a sort of "dustologist," I was asked, with considerable urgency, to investigate the matter. I visited the facility and collected samples of airborne dust only to find that the air was in fact exceedingly clean. The particles were all in the submicron range and thus were not likely to cause the problem. I also placed micro-

scope slides at various places in the room, and left them exposed overnight. When I examined the slides next morning I found they had collected lint fibers and some large plate-like particles. (By "large" I mean
relative to the dust particles. They could barely be seen with the naked
eye.) It took only a little more time to identify the fibers as coming from
the Kleenex-type tissues being used by the workers. The flat, plate-like
particles proved to be dandruff. A supply of dust-free handkerchiefs and
some simple changes in the hair covers, as well as a few suggestions about
cosmetic use (to reduce flaking from the face) and the problem was
solved! A short random walk, a day's work, and perhaps somewhere far
away bombs fell more accurately and someone died from this chance encounter.

X-rays and radium were the limits of my involvement with radioactivity during World War II. Unknown to me and my associates, however, a
giant new industry had been created under the code name Manhattan
Project to develop the atom bomb that was used against Japan in August
1945. I remember the headlines I saw for the first time at a newspaper
stand in Manhattan. There was very little information in the article, but
enough so that I could read between the lines and understand what happened. What I had no way of knowing was how that development would
affect my career.

There is much more about the war years that might be of interest, but
what I have recounted should give the flavor of the period and convey the
enormous opportunities I was given to accumulate experience that would
be valuable in later life. It was also satisfying that many new industrial
processes and materials were introduced with minimal effects on the
health of the workers. The record was by no means perfect, for we did
not yet know the extent of the lung disease that would develop in later
years among the asbestos workers. But the methods of industrial hygiene
that had developed since the First World War had proved to be effective.

Thirty years later I began to reflect on the fact that, with few exceptions, there had been no long-term followups of the cohorts—a group of
people who share a common experience within a defined time period—of
wartime industrial workers who had been exposed to toxic materials.
During the war, our primary concern was to prevent acute episodes of
occupational disease. But what were the long-range effects of Halowax,
mercury, zinc chromate, and other toxic substances to which the workers
were exposed at levels that were high by present standards? I attempted
to persuade the staff of the National Institute of Occupational Health

and Safety (NIOSH) to call a conference of the surviving occupational health specialists who had been in practice during and prior to the Second World War, in the hope that we could identify groups of workers that should be studied to provide much needed information.

NIOSH did not adopt my suggestion but they were so impressed with my account of the Halowax cases, about which they had no prior information, that they visited the two plants in New York. Much to the delight of the investigators, they found medical files on several hundred ex-workers who had developed chloracne from exposure nearly forty years earlier. An epidemiological study of those workers is now in progress and should provide important information about the delayed effects of severe exposure to chlorinated organic compounds.

It was not until 1985 that I found funds for the conference. It was sponsored by the Chemical Industry Institute of Toxicology (CIIT) and was held in March 1987 in Research Triangle Park, North Carolina. Unfortunately, the attendance included only sixteen survivors. There could have been more participants had NIOSH supported the conference five years earlier, when I first suggested that it be held. Some of our colleagues had died in the intervening years, and others became too frail to travel. However, the conference was productive, and CIIT epidemiologist Richard Levine and I summarized the proceedings and recommendations of the conference in an article published in the *Journal of Occupational Medicine*.[5] We succeeded in identifying a number of cohorts for possible future study. However, the required medical and industrial hygiene records will become increasingly hard to retrieve in the years to come.

One notable example of followup studies that were needed but not conducted was in the pharmaceutical industry, in which I developed considerable interest during the war years. Liberty Mutual provided workmen's compensation coverage for a number of the large pharmaceutical manufacturers, in whose plants workers were exposed to the dusts and fumes of potent products. Ventilation was often inadequate, and I demonstrated that the workers were inhaling pharmaceutical products in worrisome amounts each working day. In fact, in more than one plant in which female hormones were being produced, the male workers developed enlarged breasts and lost their sexual drive. Elsewhere, workers who produced atabrine, the anti-malarial drug being supplied to the troops in the field, were suffering from yellow discoloration of the skin. In one survey I found that employees who were producing a vasodilatant that also had a stimulating effect, were being given sedatives at the end of each day. Some of the pharmaceutical companies, even as late as the 1940s, seemed unaware of the need to protect their employees from the powerful drugs they inhaled and ingested.

Before closing my account of my eleven years with Liberty Mutual, I should describe one additional memorable experience. The defeat of the French armies early in the war and installation of Nazi-controlled governments in France and its African colonies left the allies with no harbors on the west coast of the African continent. Since Liberia was firmly in the allied camp, it was decided to build an artificial harbor on the Atlantic coast near Monrovia. A key feature of the harbor was to be two gigantic jetties constructed from great quantities of stone from local quarries. By the war's end the quarrying was in progress and there were indications of serious silicosis hazard to those drilling and blasting the rock. The Navy had contracted with a large U.S. construction company to build the harbor and requested that Liberty Mutual send an industrial hygienist to the site. I was given the assignment.

I originally planned to leave in early 1946 by military transport, but the trip was postponed twice and I couldn't leave until late fall, by which time commercial transatlantic service was reinstated. I booked passage on the first nonstop commercial Air France flight from New York to Paris. As I recall it, the plane carried about sixty passengers, of which all but two were missionaries returning to Africa after the turmoil of the war years. The one other non-missionary passenger was a young man on his way to the Liberian plantations of the Firestone Rubber Company for a two-year assignment. We were flying on a Lockheed Constellation, a beautiful plane that had just been placed in commercial service. Our itinerary was to fly nonstop to Paris, and then to Roberts Field, a U.S. air base in Liberia.

It was, of course, a gala occasion with excellent French champagne and cuisine abundantly available. While we were enroute the U.S. government made a decision to ground all Constellations because of a defect that required correction. Air France was not subject to the U.S. decision but wisely decided to comply with it, so that when we reached Paris we learned that it would be necessary to change planes. This would not be a problem nowadays at a major airport but this was about one year after the war ended and replacements for commercial aircraft were not readily on hand. It was necessary for us to remain in Paris for several days as guests of Air France which, in its embarrassment about the matter, was an excellent host.

I spent nearly one month in Liberia and had an excellent opportunity to explore that interesting country. I soon found that the danger of silicosis among the workers was minimal, but it was necessary for me to remain for about two weeks awaiting return transportation. During my wait I became interested in the high incidence of a particularly virulent form of malaria among the American workers. They lived and worked on the ocean front under excellent sanitary conditions, so there should not have been a malaria problem. Since this was an occupational disease insofar as the Americans were concerned, and many claims for malaria had been filed, I looked into the matter and found that there was a striking correlation between illness from malaria and gonorrhea, an endemic venereal disease. The correlation was so striking that it was not difficult to convince the construction superintendent that his men were being infected with the two diseases during their nocturnal excursions into the "bush." This was an inevitable conclusion, since there were no women in the camp. Whenever I identified a new occupational health problem it was my practice to devise a solution. In this case the problem could be eliminated by keeping the men in camp at night. I never did learn if the superintendent was successful.

While waiting for an army plane to take me to Dakar in what is now Senegal but was then French West Africa, I stayed at a comfortable officers' club at Roberts Field. There was an amiable major in charge of the field and in the course of conversation I asked him what his mission was in Liberia, where there didn't seem to be much for him to do. He replied that he and his men were there to maintain the airfield and provide the services required by U.S. military planes that stopped off from time to time. On the flight to Dakar in an army C-47, my curiosity led me to ask the pilot what *he* was doing in that part of the world. I was surprised when he replied that his mission was to bring mail and supplies to the military garrison at Roberts Field! I suppose that this kind of confusion is inevitable when demobilization is taking place after a global war.

During the months after the war ended, I again began to think about my future. It was ten years since I had received my B.S., and although I had obtained an enormous amount of experience and felt quite secure in my profession, I was becoming self-conscious about my lack of graduate education. The field of industrial hygiene was dominated by engineers, chemists and biologists, but there were very few physicians, and I came to the conclusion that it would be a good strategy to combine my ten years of intensive and varied experience with a medical degree. At that time there were no physicians on the Liberty Mutual staff who specialized in occupational disease, and the officers of the company thought it would be mutually beneficial if I did obtain an M.D. They also told me that the company would be willing to finance the cost of my medical education if I would agree to remain in its employ after I became a physician. This was an extraordinarily generous proposal, and I found it flattering as well.

It was suggested that I discuss the matter with Dr. Dwight O'Hara, who was a consultant in preventive medicine to Liberty Mutual, and Dean of the Medical School at Tufts University in Boston. He strongly endorsed the idea, although he cautioned that "people age by decades" and that I might find myself at a disadvantage in competition with classmates who were ten years younger than I. Although no decision was made, I decided to begin completing my pre-med requirements, and started to take part-time courses in the spring of 1946 with the hope that the requirements would be completed in time for me to matriculate into medical school in the fall of 1947.

However, this was not to be. In January 1947, the newly formed Atomic Energy Commission was given responsibility for developing the industrial, medical, and military applications of atomic energy. This involved a transfer of those functions from the Manhattan District, the U.S. Army organization that had been responsible for the program during the war years. As one of its first actions, the commission appointed a committee of prominent scientists and engineers—including Harvard's Philip Drinker—to tour the nuclear facilities throughout the United States and to report on the adequacy with which the potentially serious health and safety problems of the program were being managed. Among other things, the committee recommended that the AEC establish the industrial hygiene and safety capability required to deal with problems inherited from the war years. It was subsequently decided that a laboratory would be established in the New York Operations Office of the AEC to provide help for many industrial and research facilities. To staff the laboratory, Drinker sought the advice of Leslie Silverman, his junior associate at Harvard, with whom I had recently tested my welder's helmet. That coincidence resulted in Drinker's recommendation that I be offered a major position in the laboratory. Within two years I became its director. Once again my plans to resume my formal education were interrupted, this time forever.

FOUR

Atomic Energy Commission

Development of the Health and Safety Laboratory

W HEN I JOINED the staff of the Atomic Energy Commission I entered a new world in which I at first continued to be involved with occupational hazards but then moved on to more general problems of environmental protection, risk assessment, the uses of atomic energy in peace and war, congressional hearings, international relations, national research policy, and government administration.

The AEC was the postwar creation of Congress after months of debate as to whether nuclear energy programs in peacetime should be managed by a civilian or military agency. The wartime program had been limited to the development of the two types of bombs that were dropped on Hiroshima and Nagasaki in August 1945. The atomic bomb project had required construction of large centers of research, development, and production at Oak Ridge, Tennessee; Hanford, Washington; and Los Alamos, New Mexico. In addition, a network of smaller industrial facilities provided special materials such as uranium, beryllium, and thorium. There were also dozens of industrial and university laboratories with various special capabilities. This major new industrial complex constituted the infrastructure for the developing nuclear industry. How to manage it and define its mission were major questions that emerged from World War II.

The mission of the AEC, which assumed its responsibilities on January 1, 1947, was to develop and produce nuclear weapons on a schedule determined by the president and to develop a program for the application of nuclear energy to civilian needs. These included the generation of electrical energy, the production of radioisotopes for use in research, industry, and medicine, and the development of extensive programs of biological and physical research. Under the AEC it was intended that the government would limit itself to developing the program policies and the research and production goals, but would have the various operating functions executed by contracts with the private sector. This, in fact, was how the wartime program had operated: The large centers at Oak Ridge, Los Alamos, and Hanford were actually staffed and operated under contracts with universities and industrial companies.

The New York Operations Office (NYOO) of the AEC (originally known as the Office of New York Directed Operations) had been an important headquarters of the wartime program and many of its staff were among the first recruited when the Manhattan Project was formed. NYOO was one of five regional offices established by the AEC. However, in addition to responsibility for the administration of AEC contracts in the eleven northeastern states, it was also charged with administering all AEC procurement of raw materials such as uranium, thorium, beryllium, and, in the beginning, heavy water.

Bernard Wolf, a young radiologist who had recently been discharged from the U.S. Army Medical Corps, had been assigned to Oak Ridge during the war. He had agreed to join the AEC staff in New York to establish a medical division that would serve as a service and trouble-shooting unit for NYOO. The division would have a staff of industrial hygienists and health physicists to assist the many universities and industrial companies under contract with NYOO that did not have the expertise developed at the large research and production centers such as Oak Ridge, Los Alamos, and Hanford. For example, the companies that mined and concentrated uranium ore, extracted the metal, refined it, and then machined and fabricated the reactor fuel were relatively small and could not be expected to have the expertise and specialized equipment to understand and control their occupational health problems. The need for the NYOO Medical Division was identified by the committee of which Philip Drinker was a member. When it was formed, the Medical Division was the only laboratory in the AEC staffed by government employees. All other laboratories, as well as research and production facilities, were operated by universities or industrial companies under contract with the commission.

When I met Wolf, he explained what was being planned and told me that Drinker had said I was someone who could assist him in developing and directing his division which then consisted only of a secretary and administrative assistant. For the establishment of a laboratory, Harris D. LeVine, an electrical engineer, was directing construction of the electronics and machine shops as a consultant to Wolf, pending the granting of his security clearance, which took about four months. Laboratories for physical and chemical measurements would also be established. Wolf's immediate need was for someone to organize and direct the laboratories and field staffs. It took me no time at all to decide that I was interested in the position. I left Wolf with the understanding that he was interviewing others and would be in touch with me when he made his selection.

I realized that my interview went very well, largely due to the remarkable "chemistry" that existed between Wolf and myself. We thought alike and communicated easily from the beginning. I was also aided in the interview by the fact that when President Truman announced in August 1945 that the bombs had been developed and used against Japan, I had begun a program of reading on radiological physics and the health effects of radiation. By the time of my first meeting with Wolf I had read much of what was available in the open literature.

A few weeks later I was offered and accepted the position over the telephone, subject to my receiving the required security clearance (called "Q"). That was in the summer of 1947, and during the four month waiting period I read everything I could find that I thought would be helpful to me. During this time I saw quite a bit of Wolf, who assisted my orientation by giving me unclassified material to read and by introducing me to many of the people in the New York area with whom I would have future relationships. I also assisted in planning the laboratories, which were to be located in a building on the site of what is now Lincoln Center, but was then a sturdy automobile warehouse that had been taken over by the government.

The clearance procedures were undertaken by the FBI in thorough fashion. I was required to fill out a lengthy form in which I recorded information about myself, my wife, and our parents and grandparents. The bureau then interviewed many associates, but not my family or myself. It was an amusing period for me because I had told only my wife, my parents, and my immediate supervisor at Liberty that I was a candidate for a position with the AEC. Our neighbors and many of my office associates were mystified by visits from FBI agents who were not permitted to explain why I was being investigated, but who inquired about my personal habits, my political inclinations, the magazines to which I subscribed, and the organizations to which I belonged. Many of the individuals who were interviewed became concerned about me and thought I should be told that the FBI was interested in my past! This some of them communicated to me in hushed tones, only to have their anxiety relieved when my clearance was issued and I announced that I was joining the AEC staff in November 1947.

During the four-month waiting period I had learned that there were two urgent matters that required immediate attention. The AEC had a need for large amounts of two metals, beryllium and uranium, and there were urgent health problems in both industries. The problems of the beryllium industry were particularly acute. However, it was also essential to assemble the staff and provide laboratory facilities. In the months ahead I divided my time between the needs of the laboratory and the problems in the field.

STAFFING THE LABORATORY

Although the beryllium studies in particular kept me in the field much of the time, it was essential for me to be at the laboratory at every opportunity to complete the necessary recruitment of personnel and participate in decisions about the laboratory layout and the major equipment needed. Never did anyone have an easier time in such crucial matters. The AEC was a new government agency with unrivalled glamour. It was the beginning of an era of respectability for advanced technology, and atomic energy held the field almost alone. Space exploration was still ten years away. Many of the scientists and engineers who were involved in the work of the Manhattan District were eager to remain with the new commission, and those who had been demobilized from other wartime activities wanted to join its staff and laboratories. In addition, the Congress wrote the Atomic Energy Act in such a way that the AEC employees were given all the advantages of civil service employment but none of the disadvantages. Applicants could be hired on the basis of their qualifications without the need to work their way up civil service listings. There seemed to be unlimited money for equipment, and the AEC had first preference on all lists of excess equipment, much of it still in packing crates, that was left over from the war. Except for the need to wait the four months for security clearances, there were no obstacles to establishing a first-rate laboratory, and we took full advantage of our favorable circumstances.

When I arrived, Harris D. LeVine was putting together the laboratory and staff needed for the design and development of the required specialized equipment. Hanson Blatz, a radiological physicist, joined us to assemble a staff of health physicists and the support facilities they would require. William Harris, a well-known industrial hygienist from the New York State Labor Department, joined our staff as chief of our industrial hygiene section. John Harley came to us from Rensselaer Polytechnic Institute to establish the analytical chemistry laboratory. We needed a safety engineer and succeeded in attracting Edward Kehoe from his position in the New York City Fire Department. Bernard Wolf, recognizing with characteristic wisdom that we would face complicated statistical problems, was instrumental in recruiting A. E. Brandt, who had recently retired from the Naval Research Laboratory. Except for Brandt. who was in his fifties, we were a young group, all around thirty-five. We stayed together for about a dozen years during which time we assembled teams of younger men and women who became leaders in their fields. Wolf contributed enormously to the laboratory development, but left after two years to return to the practice of medicine as Chairman of the Department of Radiology at the Mt. Sinai Medical Center in New York.

When Wolf departed in 1949, we were still called the Medical Division, although he had been the only physician on a staff of about fifty, mainly physicists, engineers, and chemists. It seemed for a while that the AEC would replace him with another M.D. and a number were interviewed for the position. I participated very unenthusiastically in these interviews because I didn't see why we needed a physician as director, and I knew it would be difficult to find a physician with the varied interests and experience required. This is not so today, but in 1949 most physicians working in the field of occupational health were concerned with accidental injuries. They knew little about industrial hygiene and nothing about radiation safety.

It was about that time that the Washington headquarters of the AEC was organizing a Division of Biology and Medicine, under the part-time direction of Dr. Shields Warren, a distinguished pathologist from the Harvard Medical School. Warren was a naval medical officer who had become involved with the Manhattan District at the end of the war and had led a group of specialists into Nagasaki to survey conditions shortly after the Japanese surrender. (By coincidence, the survey of conditions in Hiroshima was studied by a team led by Stafford Warren, a radiologist who was in charge of health matters for the Manhattan District. There was often confusion between the two "S. Warrens.") Shields Warren was being assisted temporarily at AEC by Lauriston Taylor, a well-known radiological physicist who had long been on the staff of the U.S. Bureau of Standards. Warren saw no need for the NYOO Medical Division, except to help clean up conditions in the beryllium and uranium plants. Once that task was done, Warren believed the staffs of the national laboratories being established by the AEC could be called upon whenever field assistance was required. Since he expected that the Medical Division would be disbanded soon, Warren saw no reason why I should not succeed Wolf as director. That decision was made during a visit to Warren's Washington office with Wolf. Although I was happy to accept the directorship if only for a short while, it was not easy to conceal my disappointment over the fact that I would be presiding over its dissolution. As it turned out, however, our group was so useful to the AEC and Warren's Division of Biology and Medicine, that in subsequent years the laboratory was greatly expanded in size and given increased responsibilities. Shields Warren and Lauriston Taylor (who soon returned to his position at the Bureau of Standards) both became enthusiastic supporters of our program.

Warren's opposition to our laboratory was largely the result of his lack of familiarity with the practical requirements of industrial hygiene. His orientation was towards medical science, not the applications of the physical sciences to the solution of medical problems in industry. He was not acquainted with the work of industrial hygienists and had no familiarity with the complex network of plants and laboratories for which the AEC had responsibility. In addition, the NYOO was under the administrative direction of the AEC Production Division, and Warren couldn't understand why that division should be involved with matters that he considered the responsibility of the Division of Biology and Medicine. I am certain that another element in the decision was the fact that it had become AEC policy from the start to maintain a small staff of government employees who would administer the research and industrial work of the commission through contractors. The AEC was not expected to have "hands on" people on its staff.

Based on the decision made in Washington, the work of the Medical Division could continue for a while. I was the new director, but the name was changed to the Health and Safety Division, which I believed more accurately reflected its function. After a while the word division was changed to laboratory, and for many years the organization was known by the acronym HASL. When the functions of the AEC were incorporated into the Department of Energy in the late 1970s, the name was changed to the Environmental Measurements Laboratory (EML). What we didn't realize during the Washington conference was that HASL would be an extremely useful trouble-shooting arm of the commission, and would pull many of the hottest chestnuts out of the fire for many years.

One immediate problem that developed because of Wolf's departure was that I needed a physician as a staff assistant. Wolf had been the only physician in the group even though we had a continuing need for guidance and advice on medical matters. I first searched for a physician with industrial medical experience but soon found that, as in the search for a replacement for Wolf, industrial physicians with the necessary training and interests in the basic sciences were not available. At the suggestion of Norton Nelson of NYU I contacted Roy Albert, who was finishing a fellowship in cardiology and had been associated with Nelson at the Fort Knox physiological laboratories during the war. Albert agreed to join the HASL staff as Medical Officer at the beginning of a distinguished career in radiobiology. This was the start of a long association between the two of us. In 1959 we would both join the faculty at the New York University Institute of Industrial Medicine (later Environmental Medicine) where we would be colleagues for about twenty-five years.

At Warren's request, Albert transferred to AEC headquarters after about two years and was replaced by Joseph Quigley, a physician in the Du Pont industrial medical department.

HEALTH PROBLEMS IN THE BERYLLIUM INDUSTRY

The metal beryllium was discovered in 1797 and was originally called "glucinium" because of the sweet taste of some of its compounds. It is the principal element in aquamarines and emeralds, both of which are forms of the mineral beryl, one of the important ores from which beryllium is obtained. It is the lightest of the metals, with chemical characteristics similar to calcium and magnesium. From this similarity to nontoxic metals, one might conclude that it should not be a toxic substance but, to the contrary, it is among the most dangerous of all the metals.

Beryllium has many useful properties. When alloyed with copper, it produces a remarkably elastic metal which is used in springs and similar products. The oxide of beryllium is so refractory that it makes an excellent furnace lining. The metal itself is so light and so strong that it has many uses in the aerospace industry. Zinc beryllium silicate was an ingredient of the phosphors in fluorescent lamps until 1949 when it was replaced by a substitute material.

The element also has desirable nuclear properties. It is virtually transparent to X-rays, and is thus useful for windows in X-ray tubes. It becomes a compact source of neutrons when mixed with an alpha-emitting nuclide such as polonium 210. It is such a light element that it is an efficient neutron moderator, which makes it useful in reactor design. For all of these reasons, beryllium was an important metal to the AEC.

A few cases of a disease similar to pneumonia were reported among workers in European beryllium extraction plants in the 1930s, but this was thought to be due to the acid salts of the metal rather than to beryllium itself. Three similar cases were also reported from an extraction plant in Ohio in 1943[1] but these cases had attracted little attention. By coincidence, a few cases that were originally diagnosed as a chronic lung disease known as Boeck's sarcoid were reported from a tuberculosis sanatorium in Massachusetts. These cases attracted attention because they were all women who had been working in a fluorescent lamp plant in which beryllium was a constituent of the phosphor. The chronic disease seen in Massachusetts seemed unrelated clinically to the acute pneumonia-like symptoms seen in Ohio, but the fact that the two seemingly distinct types of diseases resulted from exposure to the same metal added to the mystery.

The number of cases reported from Ohio and Massachusetts increased rapidly between 1943 and 1947, by which time it was realized that a major new occupational health problem existed. Even so, there was not yet general agreement that the problem was due to beryllium. The United States Public Health Service (USPHS) failed to demonstrate toxicity in experimental animals exposed to beryllium, and its scientists therefore did not agree that beryllium was toxic. Studies of workers in Pennsylvania also showed a high incidence of the acute pneumonia-like disease seen in Ohio, but the director of the state's Division of Industrial Hygiene was influenced by the position taken by USPHS and refused to attribute the cases to beryllium exposure.

By 1947 cases of both the acute and chronic forms of the disease were also being reported from AEC laboratories as well as from lamp plants other than the one in Massachusetts.

The first major conference on beryllium disease took place at the Trudeau laboratories in Saranac Lake, N.Y. I was fortunate that it coincided with the start of my association with the AEC. Upon my return from the week-long conference, we launched a program to give us an understanding of the factors that were responsible for the disease or diseases being reported. The University of Rochester, which had undertaken animal studies on materials of interest to the wartime atomic energy project, already had been conducting animal experiments involving beryllium; Wolf requested that it accelerate its research. The Columbia University School of Public Health was requested to begin case finding in the communities in which the three extraction plants in the United States were located. Dr. Leonard Goldwater and his associates vigorously sought and located numerous additional cases. Many cases of the acute form of beryllium disease had been misdiagnosed as pneumonia unrelated to the patient's occupation. A number of these cases had been fatal.

The Columbia investigators also found that a few chronic cases had occurred but were misdiagnosed originally as Boeck's sarcoid. Most alarming was the finding that the chronic form of beryllium disease had occurred among residents near the production plants in Lorain, Ohio and Reading, Pennsylvania. This was an unusual development, since cases of human metal poisoning among the residents near industrial plants were unknown then and are extremely rare even now.

The field studies that correlated the occurrence of disease with environmental factors were performed under my supervision by the newly formed industrial hygiene section of the NYOO Medical Division. We were a small group only beginning to organize. Three of us, Alfred Breslin, Christian Berghout, and myself, spent several weeks in the field gathering data. Because our chemistry laboratory was not yet in operation, we arranged with the University of Rochester to analyze the samples of atmospheric dust, urine, and human tissue that we collected.

Most of our environmental research was done in two Ohio communities near Cleveland, Lorain and Painesville, where two extraction plants were located. Both plants were having serious health problems so the managements were extremely cooperative. A third plant near Reading was also having a serious problem but the management refused to acknowledge that it was due to beryllium. This position was supported by the director of the state Division of Industrial Hygiene who, as noted earlier, was conforming to the position taken by the U.S. Public Health Service.

The facts known to us when we started our studies in the fall of 1947 can be summarized as follows:

1. The chronic form of beryllium disease was occurring among workers in the fluorescent lamp plants and had also been reported in foundries in which molten beryllium-copper alloy was being cast, as well as in laboratories in which beryllium metal was being machined. A few cases of chronic disease had occurred among workers in the Lorain, Ohio plant and, even more alarming, there were cases among residents in the neighborhood. (One case of chronic disease had also developed in a woman who lived across the street from a fluorescent lamp manufacturing plant.)
2. The only acute cases in the lamp industry were from a plant in which the phosphors were being prepared, not in the lamp manufacturing plants themselves.
3. In the plant located in Painesville, Ohio in which beryllium oxide was being produced, several deaths had occurred from the acute disease among workers exposed to the oxide, but no chronic cases had been reported.
4. In the largest of the plants we were investigating, that in Lorain, there were a large number of acute cases reported among workers exposed to the acid salts of beryllium, but not among those exposed to the beryllium oxide being produced at that plant.
5. The acute disease had not been produced in experimental animals exposed to the oxide. This was puzzling because of the Painesville experience, but was consistent with the experience at Lorain. The chronic disease had not yet been produced in any species of experimental animal.

Because of the virulence of the acute form of the disease, our first objective was to develop enough of an understanding of it to permit the setting of standards of safe practice designed specifically for controlling the acute disease.

The Acute Disease

By the time we began our studies it was recognized that the acute disease was a chemical pneumonitis similar to that seen among workers exposed to high concentrations of nitric oxide fumes, or other chemicals that acted as irritants to the lower parts of the lung. The pneumonitis ran its course in a few weeks, and if the worker recovered there were no apparent aftereffects.

We carefully investigated the histories of workers who had developed beryllium pneumonitis and found that in many cases there was evidence of accidental exposure to massive amounts of airborne beryllium compounds. We thus developed the hypothesis that the pneumonitis was the result of relatively high short-term exposure associated with accidents.

We tested this hypothesis when, as an associate and I were collecting air samples at the Lorain plant, a gasket on a furnace failed and allowed copious quantities of beryllium fluoride fumes to fill the air of the room in which we were collecting our samples. The eight men who were working in the room at the time were placed under medical surveillance; one of them developed a mild case of pneumonitis two days later. Additional information led us to the conclusion that the acute disease would not develop if the concentration of beryllium could be kept below twenty-five micrograms per cubic meter of air, even for short periods. We recognized that a lower limit was no doubt needed for control of the chronic disease.[2]

However, we were faced with a major mystery. None of the workers exposed to beryllium oxide at the Lorain plant had developed the acute disease despite the fact that the concentrations were much higher than twenty-five micrograms per cubic meter. In contrast, many of the Painesville workers who had been affected were less severely exposed than those at Lorain. The chemical form of the two oxides was identical, but there was a difference in the particle size because of differences in the temperature at which the oxides were produced. We thus developed the hypothesis that if the temperature of the final step in the oxide production process could be increased, there would be no more cases due to oxide exposure.

We had been investigating the working histories of acute cases in the U.S. from exposure to beryllium oxide and found that all but one had been exposed to the oxide produced in Painesville. Our hypothesis thus did not seem to stand up under the test of field experience, since there was one exception, at a Westinghouse plant in which we were told that only the oxide produced in Lorain had ever been used. I mentioned this to the Westinghouse industrial hygienist and fortunately he took the time to reexamine the matter. He apparently took some of the purchasing records home with him one weekend, because he called me excitedly on a

Saturday night to tell me that there was a short period when Lorain couldn't supply the oxide and it was provided to Westinghouse by the Painesville plant. It was during this period that the case in question developed.

I had already contacted the investigators at the University of Rochester to determine which of the oxides they had used in the animal experiments in which they had been unable to produce the acute disease in rats. I learned that they had been exposing the animals to the Lorain oxide. They repeated the experiment using Painesville oxide and quickly demonstrated that the pneumonitis could be reproduced.[3]

All of our studies of the acute disease were completed in four months, in time to present our findings at the annual meeting of the American Industrial Hygiene Association in April 1948. The standard we recommended has now been adopted throughout the world. The last fatal case of beryllium pneumonitis occurred in late 1947, and there have been no known acute cases in the U.S. in about twenty-five years.[4]

The Neighborhood Cases

The first known case of chronic beryllium disease (now known as berylliosis) in Ohio was reported by a physician who had originally diagnosed the cause of death in a woman who lived across the street from the Lorain plant as Boeck's sarcoid. A year or two after the woman's death, when it became known that a sarcoidlike disease was occurring among beryllium workers, the physician reviewed the case and concluded that the woman had actually died of berylliosis. Other cases were soon found in the neighborhood. It fell on me to call on the chairman of the Lorain company to inform him of our findings. It was an emotional interview, in which a sensitive executive was faced with what may have been the first time an industrial company was responsible for a chronic and fatal lung disease among persons living near one of its plants. He understood immediately that it was necessary to determine how many cases existed in the community and asked how this could be done. When I suggested that there should be a mass X-ray of the residents, he requested that I make the arrangements for him. Accordingly, Bernard Wolf and I visited the Ohio Commissioner of Health to explain the situation to him and to request that such a survey be undertaken by his department. About six thousand residents were X-rayed in June 1948. Eleven cases were found to exist. All but one case lived within three-quarters of a mile from the plant, and these were assumed to be due to pollution of the air. The exception was a woman who lived a considerable distance from the plant and was different from the others in that her husband worked in the plant and she was exposed to the dust from his clothing. This we demonstrated by measuring the amount of dust generated by clothes washing, using the identical procedures she employed. The chore of washing soiled work clothes resulted in even greater exposure to beryllium dust than the air pollution close to the plant.

We undertook an extensive program of air sampling in the neighborhood and by consulting the company production records, we could estimate how the air pollution varied during the operating history of the plant. We demonstrated a correlation between the estimated levels of air pollution and the incidence of berylliosis within three-quarters of a mile from the plant and concluded that the disease would be prevented if the concentration of beryllium in the air could be maintained below 0.01 micrograms per cubic meter of air.[5] This standard, recommended forty years ago, is also now accepted worldwide. It was the first standard adopted for the control of air pollution from industrial sources, preceding by more than twenty years the establishment of other air pollution control standards such as those for lead, sulfur dioxide, and carbon monoxide.

It was of course also recommended that the beryllium workers be provided with work clothes at the plant, and that no contaminated apparel be taken home.

Workplace Standards

The epidemiology of berylliosis among workers in the plants and laboratories that used beryllium was confusing at the time of our studies and remains so today. It has not been possible to establish a relationship between the occurrence of disease and the severity of exposure, which of course was done with the neighborhood cases. The standard established for community air could have been extended to the workplace, but this would have been unnecessarily strict. No plant could have complied with so low a standard and the beryllium industry would have shut down. Moreover, there were good reasons why the workplace standard should be higher than the standard applicable to the general public. The worker is exposed for only eight hours a day for five days a week, whereas people living in the neighborhood may be exposed continuously, 24 hours a day, 365 days a year. Also, workers are subject to medical supervision that is more likely to result in early diagnosis should signs of disease begin to develop. Other reasons could be cited, such as the fact that members of the general community are exposed from birth, in contrast to workers who start exposure as adults.

In the absence of an épidemiological basis for establishing a standard, it was decided to assume that, atom for atom, beryllium was as toxic as the most toxic metals, such as mercury, lead, and arsenic. However, these are heavy metals, whereas beryllium is very light. If allowance is made for differences in the atomic weights of the metals, the beryllium standard would be much lower than the standards for the heavy metals. We had been discussing this matter for some weeks when the time came for a decision because a new laboratory which would use beryllium was in the final stages of design on Long Island. We had decided that the standard would be tentatively set somewhere between two and five micrograms per cubic meter. One morning I was riding to the new laboratory by taxi with Dr. Willard Machle. He was a medical consultant to the company that was building the laboratory, and had been involved with beryllium disease almost from the beginning. We knew that when we arrived we would be expected to provide the laboratory designers with design criteria and decided that a tentative MPC should be two micrograms per cubic meter. In view of the circumstances, this standard has been dubbed

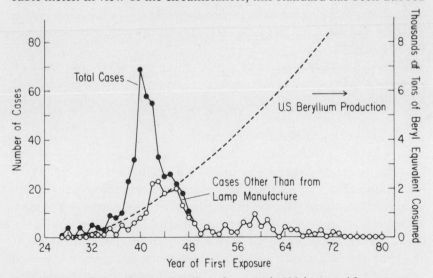

Fig. 1. Number of cases of chronic beryllium disease in the U.S. by year of first exposure, 1924–1980. Also shown is the great increase in beryllium production. The workplace standard was established in 1950.

the "taxicab standard" in recognition of the seemingly flimsy basis on which it was established. Nevertheless it seems to have been effective in controlling the disease, as can be seen from figure 1. This is all the more remarkable, considering the steady increase in the production of beryllium, which is also shown in the figure.

We had moved very rapidly to establish the three standards and felt very insecure about them. Because our research had been accomplished in less than six months, it had not yet received the peer review that normally accompanies journal publication. I recommended to Wolf that he establish an advisory panel to review our work. This was done, and the panel supported our recommendations. Accordingly, the standards were adopted by the AEC subject to the condition that they would be automatically revoked at the end of one year unless the panel recommended that they be renewed. This was done for a few years, after which the committee was dissolved because no more information was forthcoming. I am amazed that the standards we set were adopted worldwide and are still being used.

A few years later, in the introduction to a book, *The Metal Beryllium,* the chairman and president of Brush Wellman wrote, "... the New York Operations Office of the Atomic Energy Commission made essential contributions to the establishment of controls considered necessary for the safe development of the industry. Their advice has been proven practical and has given the assurance without which the industry could not advance."[6]

The beryllium experience was a good example of what can be accomplished when industry, academia, and government cooperate in dealing with environmental problems.

The Epidemiology of Beryllium Disease

The AEC made the decision to establish a beryllium case registry, which has proved valuable over the years as a means of keeping track of the numbers of cases that were developing and their environmental and medical aspects. The first registry was established by Goldwater at Columbia, but it was later expanded under contract with Harriet Hardy, first at MIT and later at the Massachusetts General Hospital, where it remained for many years until it was taken over by the National Institute for Occupational Safety and Health in Cincinnati. The case registry has made it possible to understand the epidemiology of beryllium disease in a way that would not have otherwise been possible.

The epidemiology of the chronic form of beryllium disease was quickly found to have a number of unusual features that did not conform to the pattern normally characteristic of toxic substances:

1. The employees working in the Lorain plant were exposed to dust concentrations about a thousand times greater than people living in the neighborhood. Yet the incidence of disease was about the same (1.0% compared to 1.3%).
2. The incidence of berylliosis in fluorescent lamp manufacturing was about the same as in beryllium extraction, despite the fact that the exposure was very much less.
3. Several cases had developed among secretaries in plants and laboratories that used beryllium. This was extraordinary. I knew of no other industry in which secretaries developed disease from exposure to toxic substances in their places of employment.
4. There was evidence that the disease was occurring mainly among workers who were exposed for only a short period, left work with beryllium, and then developed the disease many years later. One normally expects the incidence of disease to correlate with duration of employment. Among the beryllium workers, the reverse was true.

I was greatly troubled by these anomalies, and in 1950 I discussed them with James Sterner who had been a consultant to the AEC, but whose full-time position was that of medical director of the Eastman Kodak Company. I had prepared a summary of the epidemiology of beryllium disease but was reluctant to publish it because I could not explain the peculiarities I had found. Sterner became fascinated with the subject and after some study concluded that beryllium produced allergic reactions in certain sensitive individuals. This resulted in a joint publication which was well received by our colleagues.[7] The validity of his interpretation has since been demonstrated by studies in humans and animals. It was the first report of a chronic lung disease that resulted from an allergenic response to an inhaled metal. Our paper was awarded a prize by the Industrial Medical Association for being the most important paper of the year on the subject of occupational health.

Although my career has zigzagged many times since the beryllium studies were conducted 40 years ago, I have continued to have an interest in the subject, and recently took the time to write an update of the paper that Sterner and I published in 1951.[8] It was gratifying to find that our original conclusions had stood the test of time.

URANIUM PRODUCTION HAZARDS

The mission of the Manhattan Project required that uranium be produced in sufficient quantity, on time, and in sufficiently pure form to permit its use for nuclear purposes. The latter requirement was every bit as important as the others, because the slightest amounts of certain impurities interfered with the fission process on which both nuclear weapons and nuclear reactors depend.

When the Manhattan District was formed in 1942, the main known sources of uranium were the western U.S., Canada, and the Belgian Congo, now known as Zaire. The U.S. ore was in the form of carnotite, which had long been used as a source of vanadium. The first gram of radium (which is present in uranium ore because it is formed by the decay of uranium) was extracted by Madame Curie from Colorado carnotite. The Canadian ore was of higher grade but was located in the far north from which it could be transported only with great difficulty. The highest grade ore was from Belgian Congo, where the Union Minière de Haute Katanga owned rich deposits. The chairman of that company was apparently sufficiently aware of the importance of uranium to fear that the Congo ore would fall into the hands of the Germans. He quietly transported a large quantity to a warehouse in New Jersey, from which it became available to the project early in the war.

In 1947 the uranium was being extracted and purified in a network of about a dozen facilities, starting with an ore-crushing and blending plant in Middlesex, N.J. The ore was then shipped to plants in Niagara Falls, St. Louis, and Cleveland, for processing into purified oxides, metal, or uranium hexafluoride. The pure metal went to Hanford for fabrication into fuel elements for the plutonium production reactors, and the hexafluoride went to Oak Ridge for enrichment into the isotope U-235 by the gaseous diffusion process. Uranium is only mildly radioactive and little attention was given to the potential health problems associated with its production during the war years. It was apparently thought that the risks were slight, particularly since the plants would be operated for only a short time.

In our initial inspections we found conditions in the chain of uranium plants highly unsatisfactory because of the high concentrations of dust to which the workers were exposed, as well as to more subtle factors due to the physics and chemistry of uranium processing. The greatest problems were associated with the processing of the Congo ore, which was almost pure uranium oxide with all of the radioactive daughters of uranium, including radium and radon. The workers were exposed to direct radiation from the ore and its concentrates, as well as to the accumulations of radon and radioactive dust. In the Cleveland plant that produced uranium hexafluoride, a problem existed because the compound is a corrosive gas at room temperature and could not always be fully confined within the equipment and tanks in which it was produced. When it leaked from defective valves and gaskets it volatilized and reacted with atmospheric water vapor to produce uranium oxyfluoride, an irritating, corrosive, and toxic fume.

We collected air samples from the air of the workrooms, analyzed urine from the workers, and concluded that some of the plants were in such poor condition that conditions could not be improved except by shutting them down and building new ones. Others could be upgraded, and we made recommendations for the needed improvements.

The risk of exposure to uranium is due both to its chemical toxicity and its radioactivity. The former is particularly important when uranium is in soluble form, as was the case in the Cleveland hexafluoride production plant, where the exposures were sufficiently high to have produced minor kidney damage in a few individuals. An important question was the extent to which the uranium remains within the body as a source of internal radiation after inhalation. Today it is possible to make such measurements non-invasively by measuring the small amounts of radiation emitted from the body, but that was not possible in 1947. We made arrangements with physicians to obtain samples of tissues from workers who required surgery or who died and were autopsied. In the case of workers exposed to insoluble dust, such as the ore or the oxides, we were particularly interested in lung samples. From workers exposed to the soluble compounds such as the hexafluoride, we sought samples of kidney and bone, which are the organs in which soluble uranium tends to deposit. We did have considerable information from animal experiments performed at the University of Rochester, but we did not know how applicable the findings were to human subjects.

Samples of tissues became available from five workers over a period of four years. Analysis of the samples resulted in the finding that deposition in their bodies was much less than predicted, considering the high concentrations of uranium to which the workers had been exposed and the information available from experimental animals. In September 1955, I summarized our studies of the hazards of uranium processing at the first United Nations Conference on the Peaceful Uses of Atomic Energy in Geneva. In a paper prepared jointly with Joseph A. Quigley, we reported that uranium evidently has a low order of toxicity in man because metals such as lead, mercury, or arsenic have produced severe, if not fatal injuries at the concentrations to which the workers had been exposed.[9] At the same session of the conference, a scientist from the University of Rochester reported that, based on his animal studies, uranium is highly toxic, particularly to the kidney. We were, of course, both right, and the differences in our conclusions served to illustrate the uncertainties involved when applying animal data to humans.

In 1950, Hanson Blatz and I summarized the exposure histories of all of the workers who had been in the employ of the Mallinckrodt Chemical Company in St. Louis. This was the largest of the plants in the uranium chain and they were heavily exposed. We tabulated the doses received by each employee, as estimated from film badge measurements, dust samples, and other means.[10] More than thirty years later an epidemiological study was begun by the Oak Ridge Associated Universities of the long-term effects of their exposure on the health of the workers. When I learned that the investigators were unaware that our report existed, I brought it to their attention. It should be helpful in understanding the relationship between the severity of uranium exposure and long-term health effects.

In April 1948, Bernard Wolf and I travelled to Grand Junction, Colorado, where NYOO maintained a field office. Our purpose was to inspect the uranium mines and make recommendations for protection of the workers from the hazard of exposure to radon, which we knew could be a serious problem in mines of that kind. What we knew came from experience in the metal mines in Schneeberg and Joachimsthal in what is now East Germany and Czechoslovakia. As long ago as the sixteenth century, a German physician who called himself Agricola reported that the miners in those areas suffered from a fatal lung disease known as

"Mountain Sickness." It was not until the nineteenth century that the disease was identified as lung cancer, and not until this century that its cause was identified as the radon in the mines.

Radon is a radioactive gas that is produced from the decay of radium which, in turn, is a radioactive product of the decay of uranium. Since uranium and its family of radioactive daughter products are present everywhere in the crust of the earth, radon diffuses from the soils and rocks and is a normal constituent of the atmosphere. The European mines had been important sources of many metals for many centuries and, although it was unknown at the time of Agricola, uranium and radium were present as well, producing radon in copious amounts. In 1941, the National Council on Radiation Protection and Measurements established a standard for occupational exposure to radon, based on information available from those mines.

There was no data on the radon concentrations in the Colorado mines, and during our visit Wolf and I arranged for the collection of air samples, for medical examinations for the miners, and for other things needed in an occupational health program where there are exposures to toxic or radioactive materials. Much to our surprise, we were told by Washington that the health problems of the mines were not the responsibility of the AEC, and that they should be left to the jurisdiction of the local authorities. This was an absurd decision because the local officials had none of the special knowledge and equipment needed to deal with such matters.

Responsibility for regulating industrial safety normally resides with the individual states, but in the Atomic Energy Act of 1946 which established the AEC, Congress assigned responsibility for radiation safety in the atomic energy program to the new commission because the states did not have the specialized equipment and expertise needed to deal with the expected health problems. However, according to that act, AEC responsibility did not begin until after the ore was mined. The states thus retained responsibility for the safety of uranium mining, presumably because Congress thought the health hazards were relatively minor.

The position of the New York Operations Office was that while the act did not require that AEC be responsible for uranium mine safety, neither did it prevent the agency from doing so if it so wished. AEC would be the only customer for uranium ore. Wasn't it proper that a government agency should require that it be mined safely? The precedent had been set with beryllium, a metal that is not even radioactive. NYOO adopted the simple position that since AEC would be the major user of beryllium it should be processed safely by companies doing business under contract with the agency. The Atomic Energy Act did not specifically assign responsibility for the safety of beryllium production and use to AEC, but neither did it prevent AEC from requiring its contractors to comply with appropriate safety standards. The Washington headquarters was aware of the position taken by NYOO and not only approved the policy, but indicated that the beryllium standards established would be applied to AEC contractors countrywide.

However, beryllium production and uranium mining at AEC headquarters were the responsibilities of different divisions. Uranium procurement was under the aegis of the Raw Materials Division, which took a different view of the matter than the Division of Production, which was responsible for beryllium research and production.

Wilbur Kelley, Manager of NYOO, based on advice given to him by Wolf and me following our visit to the mines, continued to insist that if NYOO was to be responsible for ore procurement, it should retain the right to investigate the health hazards in the mines and establish safe operating criteria. A stalemate resulted, which was resolved when the Washington Division of Raw Materials removed the ore procurement responsibility from NYOO and placed it in a Raw Materials Operations Office created for the purpose within the Washington headquarters. This arrangement was a unique one within the AEC. It was a tragic decision because the state of Colorado and other western states in which uranium was being mined did little to implement their legal responsibilities for mine safety, and by the early 1960s cases of lung cancer due to radon exposure began to develop among the miners. Mine ventilation was improved, but too late to prevent an epidemic of lung cancer which would take the lives of about 500 miners.

FIVE

Atomic Energy Commission

Studies of Radioactive Fallout

THE FIRST EXPLOSION of a nuclear weapon took place on a steel tower in the New Mexican desert in July 1945, about one month before two bombs were dropped on the Japanese cities of Hiroshima and Nagasaki. The test explosion in New Mexico, known by the code name TRINITY, caused skin burns on grazing cattle several miles downwind where radioactive particles deposited on them. The radioactive cloud drifted over Iowa, where a thunderstorm washed particles of radioactive dust into a corn field, the shocks of which were subsequently shipped to a factory of the Eastman Kodak Company, which manufactured the black interleaving paper used to separate sheets of X-ray film. The radioactive particles became incorporated into the interleaving paper and caused black spots to appear on developed film that had been packaged with the contaminated black paper. Eastman Kodak described its experience in a scientific article, but it attracted little attention.

Dangerous fallout of radioactive particles occurs when a nuclear bomb is exploded so close to the ground that particles of soil and other materials are sucked into the cooling fireball as it rises. When bombs are exploded at greater altitudes, the debris takes the form of fine particles that settle to the ground, usually with rain or snow, at great distances from the explosion, after most of the short-lived radionuclides have decayed. The two bombs dropped on the Japanese cities were exploded high in the air, which tended to minimize fallout, but rain showers shortly after the explosions did result in slight, but measurable levels of fallout at both Hiroshima and Nagasaki.

The first nuclear bombs exploded after the war were at Operation

CROSSROADS, a well-publicized demonstration that took place in 1946 on the Pacific atoll of Bikini. CROSSROADS was a festive occasion attended by invited guests from many nations and a vast corps of reporters. It was apparently hoped that a demonstration of the power of nuclear weapons would have a beneficial effect on the outcome of a conference on control of atomic energy that was in progress at the recently formed United Nations.

In preparation for the tests, 162 Bikini natives were moved to another atoll. A fleet of warships, some surrendered by the Japanese and Germans and some antiquated vessels from the U.S. fleet, were anchored in the lagoon to demonstrate the effects of the new weapons on naval vessels. The first bomb was detonated high above the lagoon and the radioactive cloud drifted away without incident. The second bomb was exploded below the lagoon surface and created a massive and spectacular geyser that doused the ships with so much radioactivity that some had to be sunk. Others were towed to naval stations where they served as laboratories for development of decontamination methods. The underwater explosion caused so much contamination of the ships and lagoon that a third planned explosion was cancelled.

In July 1947, the AEC announced that it was establishing a "proving grounds" for the testing of nuclear weapons on the atoll of Eniwetok, also in the Marshall Islands. The 120 natives who lived on that atoll, like the Bikini natives in 1946, were removed to another island, thus beginning a sad saga that continues to this day. The first tests of Eniwetok took place in 1948, the year in which the Soviet Union exploded its first nuclear weapon at a test site in Siberia. This latter event took the U.S. by surprise and resulted in the acceleration of nuclear weapons development. Additional testing facilities were needed, and because of the logistic problems associated with the Pacific Test Site, a decision was made to develop a site in the United States. The site selected was near Las Vegas, Nevada, and the first exercise, Operation RANGER, began on January 27, 1951.

Preparations for the weapons tests at both the Pacific and Nevada test sites were undertaken in great secrecy by the AEC. Since NYOO was not involved, I learned about the tests as did the general public, by hearing of them on the radio. A few days after the first series of tests began in Nevada, I received a call from Harry Blair of the University of Rochester, who informed me that Eastman Kodak Company had reported that radioactive dust was accumulating on the air intake filters at their Rochester film manufacturing plant. It seems that the experience of the company after the TRINITY explosion in 1945 had alerted them to the need to take precautions against the presence of radioactive dust in the air of their many manufacturing facilities. The presence of radioactivity at the Rochester plant was reported by radiation detectors installed for that purpose. Blair also reported that the radiation background in the university laboratories was also higher than normal. The levels were not high enough for concern about their effects on human health, but they were high enough to justify the concerns of the Eastman Kodak officials about the integrity of their film-making process. The increased background was also affecting the sensitive measurements being made at the University of Rochester laboratories.

A test explosion had taken place high above the Nevada desert about thirty-six hours previously, and the radioactive cloud had drifted over the Rochester area during a snowstorm that was blanketing the Northeast. I first called the medical director at the Nevada test site and advised him of the reports we had received. He thought it was nonsensical that fallout was reported so far away. He had been to ground zero a few hours previously, and there was only insignificant residual radiation. What he failed to realize was that the bomb was detonated so high in the air that there was no local fallout, but the cloud of radioactive debris had been carried in a northeasterly direction and brought to the ground by the precipitation in the Rochester area. I immediately advised the Division of Biology and Medicine in the AEC headquarters, and was told, much to my surprise, that the Brookhaven National Laboratory on Long Island had been alerted that the tests were about to take place, and that its staff had been asked to install radiation monitoring stations "somewhere in the Northeast."

When I called Brookhaven I was told they had installed two monitoring stations, one at the laboratory and another somewhere in Maine. The instrument at the laboratory was recording nothing but the normal radiation background, which meant that that part of Long Island was not in the path of the cloud. What about the station in Maine? It was unattended in a remote area, and they would be unable to recover the information for several days! In view of the fact that there was apt to be considerable interest in the subject, I thought it was worthwhile for them to send someone to the station immediately, but I was not persuasive enough.

It had been snowing throughout the Northeast, the temperatures were below freezing, and the snow was still on the ground. It would thus be possible to measure the amount and extent of the fallout by examining samples of snow. The HASL branch chiefs, Harris, LeVine, Harley, and Blatz, conferred with me and we decided to take advantage of NYOO's many contracts with industrial companies and universities throughout the Northeast states and request that those organizations assist in collecting snow samples for analysis. We asked them to purchase quart ice cream containers, collect snow samples along designated routes, seal the containers, and deliver them by messenger to HASL over the weekend. We wanted the samples collected as soon as possible, but not until it had stopped snowing, so that we could be certain that all the precipitation would be collected. By early Saturday morning, samples were being collected in the northeastern U.S. by personnel dispatched from St. Louis, Cleveland, Rochester, Albany, New York City, and Boston. The HASL staff worked throughout Friday night assembling the equipment to evaporate the water samples for measurement in beta radiation counters, the easiest way to assay radioactivity in those days. The staff performed magnificently over that weekend and by Tuesday afternoon prepared a map showing the pattern of radioactive fallout. It was the first time this had ever been done, and Shields Warren was anxious to see the map because he was scheduled to testify at a congressional hearing and thought the committee would be interested in what we had found. I took the map to Washington that night, and Warren presented our findings to the congressional committee on the following day.

HASL had already attracted attention within the AEC for the effective way in which the beryllium studies had been conducted. The Rochester incident drove home the message that the nature of AEC activities required an in-house troubleshooting group that was responsive to the special needs of the commission whenever and wherever they might arise.

At the time of the Rochester fallout there was little appreciation of the fact that some of the nuclides present in the bomb debris were capable of being absorbed by plants and animals and could eventually find their way into food and the human body. Such information was available from studies that had been conducted at Hanford and other nuclear centers, but the information was still secret in 1951, and at HASL we didn't know it existed. What is disappointing in retrospect is that the Division of Biology and Medicine at AEC headquarters had paid so little attention to the matter that there were no advance preparations for fallout measurements. Immediately following the Rochester incident there was interest in the subject, but only because it was a scientific curiosity, not because the fallout should be investigated for its public health implications. HASL might not have been involved in further fallout studies were it not for the fact that Eastman Kodak needed information to protect its processes and requested the AEC to provide it with advance notice when a cloud from a weapons test was approaching one of their manufacturing facilities. This request was relayed by the AEC Division of Military Applications to Wilbur Kelley, manager of NYOO, with instructions for HASL to assist Kodak in any way possible.

By the spring of 1951, preparations were already well underway for tests to be conducted in Nevada that fall in two parts—BUSTER, a series of explosions above ground, and JANGLE, which consisted of a surface and an underground test. Although the latter two explosions would be relatively small, they could be expected to result in much higher levels of fallout than had the previous Nevada tests. I was invited to Los Alamos for a briefing on the proposed tests and learned that the test organization was planning to monitor the environment only to two hundred miles from the bursts. I was of the opinion that intensive monitoring should be extended to at least five hundred miles. I was told the test organization would be unable to go beyond two hundred miles, but that they would support a recommendation to monitor out to five hundred miles if HASL or others could develop a feasible plan. I stated HASL would be willing to assume monitoring responsibility for the 200–500 mile annulus around the test site if the military provided us with logistic support. It was agreed that this would be done.

We now had two responsibilities during the forthcoming tests. In addition to monitoring the 200–500 mile annulus around the test site, we would establish a network of fixed stations throughout the world.

The design of the remote monitoring system received a good deal of thought. Ideally, we would have liked to make measurements of external gamma radiation, airborne dust, and individual nuclides deposited on the ground. This was not feasible because we had a very modest budget, and the logistic problems could not have been solved if we had had the funds. Hanson Blatz came up with a practical suggestion. He had investigated commercially available adhesive films and had found gummed plastic sheets that would retain their adhesive properties when exposed out of doors in inclement weather. We ordered a quantity of one-square-foot sheets and had them placed about three feet above the ground on a simple stand that we had manufactured. With the cooperation of the U.S.

Weather Bureau, we established an initial network of forty-five stations at which the gummed films were exposed for twenty-four hours before being changed. The films were then folded, placed in an envelope, and mailed to HASL where the beta activity was measured, using an ingenious automatic counting system designed by Harry LeVine and his group. That simple sampling collection system was eventually extended worldwide and for many years provided the only available information about the levels of radioactive fallout. It was originally considered at best a simple method that could provide semiquantitative information about the passage of the clouds of weapons debris and about the places where fallout was occurring. It later developed that the data could be used to infer the rates of fallout of individual nuclides of special interest, such as iodine 131, cesium 137, and strontium 90. It is now more than thirty-five years since the gummed films were used, but the data we collected continue to attract the attention of investigators who use the information in various ways to reconstruct the radiation doses received by inhabitants in various parts of the U.S. and elsewhere.

For studies in the 200–500 mile annulus, the army detailed fifteen enlisted men to HASL in addition to two DC-3 aircraft and their crews. These men, with HASL supervisors, were based at convenient airfields and deployed to monitoring locations selected on the basis of meteorological information at the time of detonation. Our objective was to establish stations across the trajectory of the cloud, to begin sampling prior to arrival of the cloud, and to sample continuously for 36 hours. We usually established three stations following each burst. There were so many military installations near the test site that it was usually possible to establish the temporary stations at airfields attached to those bases. Each station set up gummed film stands, air samples equipped with inertial particle size sorters, and gamma radiation recording instruments. The plan worked extremely well and was repeated for each of the test series during the next five years.

For most of the period of its operation the mobile sampling program was supervised in the field by Paul Klevin, Al Breslin, John Harley, and William Harris. It was dangerous work because of the need to fly into small airfields in all kinds of weather, and I recall one tense episode in which the aircraft to which Klevin was attached was missing for half a day, which was a long time considering that the aircraft was never more than five hundred miles from its base. We sent radio messages to all air fields inquiring about the plane but received no reply because the radio operator at Hill Air Force Base in northern Utah, where Klevin had landed for purposes of monitoring, thought the plane was on a secret mission! The HASL mobile teams functioned efficiently during several series of Nevada tests during the early 1950s, and there were fortunately no mishaps.

Our studies were classified "secret" for a while but before long we were allowed to disseminate the information without restriction. John Harley and I prepared the first summary of our findings, which we published in *Science* in 1953.[1] This was the first of several articles on the subject.[2] The policy of publishing our data in the open literature was a wise one. It would have been wrong not to have done so, and from a practical point of view there was no way in which the presence of radioactivity in the environment could be kept secret. By the mid-1950s most research laboratories used sensitive radiation detectors, and many scientists were reporting increases in the background levels, which they realized was due to fallout from the weapons tests. Some of the scientists began to complain to their elected officials who made inquiries to the AEC, and these were referred by Washington to HASL because we had the needed information. Richard Nixon, Senator from California, was among those who asked for information in May 1953, when tests were in progress in Nevada. Reports of fallout began to be made even in Europe, and the distinguished British physicist Sir John Cockroft visited me to report his estimate (ten curies) of the amount of bomb debris that had deposited on the Thames valley from the tests that were conducted in the fall of 1951. The existence of radioactive fallout from the weapons tests could not, need not, and should not have been kept secret.

Although the first concerns about fallout were because of the effects on film manufacturing and radiation detection equipment, it wasn't long before many scientists raised questions about effects on human health. The studies of the fallout patterns following the surface and underground tests in the fall of 1951 led us to conclude that the effects of radioactive deposits from the explosion of nuclear weapons in time of war could be comparable to, if not greater than, the effects due to blast and heat. Although the radioactivity from the two relatively small JANGLE explosions was deposited close to the site, we concluded that the fallout from the thermonuclear weapons being contemplated could contaminate thousands of square miles of territory with lethal or near-lethal levels of radioactivity.

The sizes of nuclear explosions are described by their equivalence to the chemical explosive, TNT. A one-kiloton nuclear explosion is equivalent to the explosion of one thousand tons of TNT. The bomb exploded over Hiroshima was equivalent to about fifteen kilotons of TNT. By 1952, thermonuclear weapons equivalent to more than ten megatons (ten million tons) of TNT were being designed. This is an inconceivable quantity. Ten million tons of TNT, if placed in a pile three feet high and six feet wide, would be about 625 miles long!

Our attention turned to the contemplated explosion of the first large thermonuclear device scheduled to take place on the atoll of Eniwetok in the fall of 1952. There was much thoughtful discussion about the possible consequences of such an explosion: Some scientists believed that the force of megaton explosions would drive the radioactive dust into outer space! Our staff was less optimistic, and predicted that dangerous levels of fallout could occur for hundreds of miles downwind of the explosion. During those months I spent a considerable amount of time reading about the dust produced by volcanic eruptions, which were more violent by far than thermonuclear explosions. I was impressed by the fact that the dust was injected high into the stratosphere where it remained for years and affected the coloration of sunsets around the world. In addition to high levels of fallout that would occur within hours after an explosion in which the fireball touched the ground, it was highly likely that fallout of small particles would "dribble" from the stratosphere for many years.

Although information about the levels of fallout at the stations operated by HASL were not kept secret, almost everything else about the testing programs was classified. This included the schedules of testing, the size of the explosions, and how far off the ground they were to be exploded. In Nevada this was carried to ridiculous extremes. The tests naturally attracted considerable media attention, but the reporters and photographers were not told in advance when a test was scheduled. These took place at about weekly intervals, generally at about daybreak. Many of the reporters assigned to the testing story stayed at the gambling casinos in Las Vegas, and since they stayed up late at the tables, they would see the VIPs come through the lobbies in their field clothes, which was all the information the reporters needed to know it was time to collect their gear and head for Mount Charlestown, a picnic park from which they could have a good view of the test site, ninety miles away, but readily visible through the clear desert air.

Secrecy was much more strict at tests in the Marshall Islands, and information was dispensed only to those with the required security clearances on a "need to know basis." The trouble with the system was that one might not know what one needs to know! This was particularly true of our group because we were not part of the regular weapons testing organization and did not really understand how the classified information related to what we were doing, and therefore did not know what reports to ask for or what questions to ask.

For example, we had no appreciation of the extent of the monitoring effort that was established to detect nuclear explosions in the Soviet Union. That program was highly classified and although I was getting acquainted with some of the scientists involved, they never really said what they were doing. For all I knew, they were simply part of the large organization concerned with testing nuclear weapons. One day, after I had flown for two hours out of Kwajalein to an atoll in the Marshall Islands, as the only passenger in a Navy amphibious aircraft, I landed at the island of Ponape to install some monitoring instruments. I found a well-known chemist installing similar equipment. He was there to test the equipment he used for the remote detection of nuclear explosions in the Soviet Union. I was there to measure fallout from our own tests. Both programs were classified, and neither of us knew the details of what the other was doing. This is an example of how wasteful some security restrictions can be without contributing to the purposes for which they were established.

By the summer of 1952 the gummed-film network we established had been extended world-wide in preparation for a thermonuclear test scheduled for November 1 on the island of Engebi in Eniwetok atoll. By that time, John Bugher, who had had a long association with the Rockefeller Institute (now the Rockefeller University), had joined the AEC staff as Deputy Director of the Division of Biology and Medicine. He eventually replaced Warren as Director and became an important supporter of our laboratory. Although a distinguished physician, he was also trained as a physicist and had a full understanding and appreciation of the importance of the work we were doing. He was impressed with the value of the data that had been collected by our mobile teams in the 200–500 mile annulus around the Nevada Test Site. One day in July, only four months before the scheduled multimegaton thermonuclear explosion, he asked if I thought it would be feasible to conduct a similar exercise in the Pacific. This was something I had been thinking about for several months, and I replied that it was not only feasible, but essential that it be done. The test explosions, Operation IVY, were to be conducted by the Department of

Defense, using Joint Task Force 132, commanded by Major General Percy Clarkson. The task force was not planning extensive monitoring operations beyond the atolls in the immediate vicinity of Eniwetok. Increased surveillance was essential to better understand fallout phenomena and, above all, to protect the inhabitants of the Pacific islands. The first test of the IVY series was named MIKE. This test was of particular concern to me.

The survey methods would be far more difficult than in Nevada because the land to be monitored consisted of tiny islands scattered over thousands of square miles of ocean. However, we believed that the amount of fallout could be measured by flying low over the islands with scintillation counters. Instruments of the kind required didn't exist but they could be designed and built in our laboratory. When Bugher was with the Rockefeller Institute he had studied tropical diseases in the jungles of South America, so he had a "field man's" appreciation of the challenge that existed. Time was short, the logistics were difficult, and there were organizational hurdles to be gotten over. His eyes gleamed as he said he would try to arrange for us to monitor the Pacific islands.

With encouragement from Bugher, but without approval to proceed, we began to design and construct the instruments we needed. It would be necessary that the detector be self-contained so that it could be used in any aircraft assigned to us. It had to have a fast response time, so that changes in the radiation level could be detected from low-flying planes (200–500 feet) at normal flying speeds, and it had to have a logarithmic response so that it would be useful from 0.01 mr/hr to 100 mr/hr. In addition to the aerial measurements, we would also install land-based monitoring instruments on a few islands, but these could be of a type with which we had considerable experience, so little additional development would be required.

We received approval for our plan by the end of July and began the hectic preparations for the program. The officers of JTF-132 in Washington cooperated splendidly, and in early September I flew to the headquarters of General Clarkson at Pearl Harbor to review our final plans and confirm that the required logistic support would be made available. By that time I was high up in the civil service system, with the grade of GS-17, only one step from the top, and I found that when I travelled with the military I was given the privileges of a general officer. I found it remarkable that information about my assigned equivalent rank always preceded me and the red carpet was always rolled out. Military protocol was at a high level of efficiency.

One of the important decisions that was made while I was in Pearl Harbor was that we would be supported by CINCPAC, the headquarters of the Commander in Chief of Pacific Operations. This had the advantage of not burdening JTF-132 with providing the planes and personnel we required. In addition, the support we received from CINCPAC added prestige to our mission, and we were given space on the U.S.S. *Estes,* the flagship of JTF-132.

The instruments were completed on schedule and in mid-October we shipped several thousand pounds of equipment and supplies to Kwajalein, the atoll on which we would be based. John Harley, Melvin E. Cassidy, Al Breslin, and I followed a few days apart. I arrived in Honolulu on October 14 and conferred again with the CINCPAC staff, who seemed to have developed even more interest in our program since I had last seen them. Of special importance was their decision that all the islands of the Pacific Trust Territory should be surveyed, regardless of the fallout predictions based on meteorological forecasts. CINCPAC had reason to be concerned about the forthcoming tests. Most of the islands of the central Pacific had been Japanese possessions prior to World War II. After the war they became a trust territory, assigned by the United Nations to the U.S. for administration and protection. These functions were in turn assigned to CINCPAC with responsibility for safeguarding the indigenous population.

From Pearl Harbor I proceeded to Kwajalein which, like many of the Marshall Islands, had been the scene of violent battles only seven years before. This atoll would serve as headquarters for the HASL staff of three, who had arrived ahead of me and had already departed for the other islands on which instrumentation would be installed. I was really enjoying the VIP treatment I was receiving. I ate and slept with the admirals and generals and was extended all the privileges expected by them.

On October 25, 1952, Harley, Breslin, Cassidy, and I travelled to Eniwetok for a final test of our newly built instruments. This time we flew over the radioactive craters and debris left from the previous tests. We flew over the radioactive areas, measured the radiation levels within the aircraft, and then monitored the same places on foot. We found, as we had calculated, that the levels of radiation we were recording at 500 ft. were almost exactly one-tenth of the intensities measured at ground level. The results were quite reproducible on repeated passes over the areas measured.

On October 30 I transferred from my shore billet to the *Estes,* which was the task force command ship. I shared a comfortable room and private shower with three officers, with all of whom I was well acquainted. The next day was occupied with readiness reviews and meteorological briefings preparatory to the detonation scheduled for the early morning of November 1. The test, code-named MIKE, would be the first large thermonuclear explosion and it was expected to yield in excess of ten megatons of explosive power. There was some uncertainty about the meteorological conditions, and the final decision to proceed was not made until 2:00 A.M. on November 1, only a few hours before firing time.

Listening to the final countdown on the loudspeaker, I stood on the deck of the *Estes* as dawn approached. The deck was crowded with officers and civilians of a wide variety of backgrounds. There were congressmen, senators, top officials of several governmental agencies, and a number of prominent scientists. Sir William Penney, a British explosives expert who had been involved with the bomb project at Los Alamos during the war, stood next to me and commented about the comfortable quarters to which we had been assigned. "I must say," he said with typical British humor, "if one must join the Navy, there are advantages in being an admiral." The contrast in the various types of chitchat during that last nervous hour was remarkable. There were more than a few observers who expressed the hope that, for some reason, the thermonuclear reaction would fail. It would be a better world without thermonuclear weapons. Others were betting on the yield of the explosion. One prominent physicist was particularly somber: he had told me the night before that he could not fully rule out the possibility that the thermonuclear reaction might "ignite the atmosphere." There was a good deal of the normal gallows humor in which men indulge at tense moments.

The explosion was unexpectedly awesome. First we saw the fireball, perhaps one mile in diameter, through the dark glasses we were required to wear, and at first it didn't look very different from much smaller nuclear explosions I had seen because we were witnessing the event from a distance of thirty miles, whereas the small bombs, sometimes only one one-thousandth the size of MIKE, were seen from much smaller distances. It wasn't until we were permitted to remove our dark glasses that we appreciated the full magnitude of the event. As the fireball cooled on its way up to the stratosphere (I later learned that it reached an altitude of twenty-six miles), it entrained enormous quantities of water vapor that condensed in progressive stages to produce a great inverted ziggurat with sharply defined steps as it punched its way through successive inversion layers of the atmosphere. As in the last minutes before the explosion, there were marked differences in the behavior of the observers. Most, like myself, were dumbstruck, but many cheered. I couldn't understand how they could cheer the fact that we had witnessed the beginning of an era when wars might be fought with bombs having as much as a thousand times the destructive power of those that destroyed the two Japanese cities in World War II, only a few years before.

The plan called for me to be in the first plane back to Kwajalein, so that we could begin our surveillance of the islands. One hour after the detonation, when the head of the mushroom cloud had expanded to a diameter of about sixty miles, and was drifting to the west at twenty miles per hour, I left the *Estes* by helicopter for the aircraft carrier *Rendover*. There I boarded a plane with a forward cockpit for the pilot, and a rear seat for one passenger, but two of us, George Cowan, a chemist from Los Alamos, and myself were squeezed into it! We were catapulted from the carrier and spent the next two hours under such cramped conditions that thirty years later George insisted that his shoulder never fully recovered from the flight! For the first hour we had to fly around the enormous mushroom cloud, but the dosimeter I was carrying recorded very little radiation from it.

When we arrived at Kwajalein, the naval officers assigned to that island were standing at attention on the tarmac awaiting us. This was in part a matter of protocol, but it was also because they were eager to hear about the results of the test. We could tell them very little except that it was apparently successful, but this they already knew since they had seen the flash of the fireball nearly two hundred miles away.

Our plan called for the sweeps over the islands to begin on the first day after the test and to continue for four days. On the morning after my arrival, Cassidy and I left Kwajalein in two PBMs, sturdy long-range Navy flying boats. We were each to sweep about a thousand miles of ocean over which the cloud should have already passed. When we departed at 5:30 A.M., the radiation background at Kwajalein was about ten times its normal level, indicating that part of the cloud had passed our way. During that first day I made measurements over fifteen exquisite atolls with Polynesian names like Taongi, Utirik, Wotho, and Bikini. The weather was very bad, and there was no way to avoid frequent squalls that interfered with our visibility and made the flight uncomfortable. From time to time we would fly through a piece of the cloud and particles of radioactive dust would impact on the leading surfaces of the aircraft, which would increase the radiation background within the aircraft. We found that the particles would wash away when we flew through a squall. I found only minor levels of radiation on the atolls despite the fact that I had chosen the sector which I thought would receive the heaviest fallout. I thought that perhaps I had been wrong to believe that heavy fallout would occur. I did detect elevated radiation levels on a few atolls, but the highest level at the surface was only 0.6 mr/hr, which is much lower than we had seen in many parts of Utah and Nevada after tests near Las Vegas.

We realized that the islands were so far apart that significant fallout might miss them, and this was what had happened. As a matter of fact, there were a number of times on that flight when my instrument reading began to drift upscale. I interpreted these readings as the result of instabilities in the instrument, which was only recently off the drawing board. The instrument was in fact behaving properly and was recording radiation from fallout in the water. It had not occurred to me that the radioactive materials would remain near the surface long enough to be detectable, even with the effect of shielding by the water.

On November 6, after four days of essentially negative results, I flew to Guam in the western Pacific to join John Harley, who had been monitoring the islands of the Mariana and Caroline groups. The final sweep of our program was a flight from Guam to Tokyo in which we flew low over many volcanic islands, including Iwo Jima, on which one of the fiercest battles of World War II had been fought.

We ended our surveys of the Pacific islands without finding significant amounts of fallout. CINCPAC was delighted with the negative results. However, there would be other multimegaton tests in the future, and I was of the opinion that the good fortune and careful meteorological control that had resulted in minimal fallout following MIKE might not always be the case. Improved monitoring methods were needed to obtain the knowledge we needed to better understand the dangers of fallout.

I spent much of 1953 at conferences on the subject of weapons fallout, and how to do a better job of documenting the phenomenon. In the spring of that year, during the tests called UPSHOT-KNOTHOLE in Nevada, there was another occurrence similar to the Rochester fallout during RANGER. This time it was centered on Troy, New York, where it was detected by Dr. Herbert Clark, John Harley's former mentor at Rensselaer. The fallout was associated with heavy thunderstorm activity in the area and resulted in ambient gamma radiation levels of about one mr/hr. Unlike the Rochester incident, the fallout at Troy received attention from public officials, but in telephone conversations with the commissioner of health of New York State I found no sense of alarm, but only an interest in obtaining information. Professor Clark estimated that the gamma exposure to the residents from the deposited radionuclides was about 100 milliroentgen, which is about the dose received from natural sources in one year. John Harley and I flew up to define the extent of the fallout with our scintillometers but were turned back by heavy thunderstorms in the area. It wasn't until May 4, six days after the initial report, that the weather cleared sufficiently to allow a flight. Harley found that the city of Albany had accumulated even more fallout than Troy, and the ground level radiation levels averaged about 0.4 mr/hr. despite the fact that by this time much of the radioactive deposit must have been washed away by the heavy rains.

The fallout incidents were beginning to attract considerable media attention, especially in the southwestern states, where many scientists were becoming concerned about the levels of exposure being received by the general public. The photographic industry was also increasingly concerned and in late May the National Photographic Manufacturers Association sent a formal protest to AEC about the problems created in their industry by the accelerated testing schedule.

WEAPONS TESTING AND STRONTIUM 90

In the spring of 1953 I learned about Project GABRIEL, which had been underway since 1949 in great secrecy, and had as its purpose the calculation of the number of nuclear weapons that could be exploded in a nuclear war before the world would be so contaminated with fallout that long-lived effects of radiation would develop. It was then believed that there was a "threshold dose," below which no delayed effects of radiation would occur. This is no longer believed to be true, at least so far as cancer and genetic effects are concerned, but until the late 1950s the concept of threshold dose prevailed.

The last report of Project GABRIEL had been issued in 1951 and concluded that the critical radionuclide in fallout was Strontium 90 (Sr-90) which has properties similar to radium in that it deposits in the skeleton, from which it is eliminated very slowly. The limit for "threshold lethality" was estimated at between forty thousand and forty million megatons of fission equivalent. The great uncertainty implicit in the thousandfold range between the upper and lower limits was a reflection of the ignorance that existed about the manner in which the dust from nuclear bomb explosions was transported in the atmosphere, the rates of surface deposition, and the kinetics of Sr-90 uptake by plants, animals, and humans. Even the lower limit seemed huge in 1951, when there were very few nuclear weapons, but a desire to narrow the uncertainty in the GABRIEL estimate began to develop in 1953 because of the experiences with weapons testing, much of which had been documented by the HASL research.

The responsibility for reinvestigating the matter was given to the RAND Corporation, a "think tank" established by the Air Force after World War II. That organization assembled a group of about fifty scientists at their headquarters in Santa Monica for a series of discussions of the matter in July 1953.[3] I attended and was pleased that most of the available data for the discussions came from the HASL reports. The meeting was a turning point in our understanding of the hazards of fallout because it was attended by many prominent scientists who recognized that far more information was required about the physical, chemical, and biological aspects of weapons fallout. Out of the meeting came a new project codenamed SUNSHINE, after the excellent weather we enjoyed while in California. This was a most unfortunate selection that inadvertently gave a much more cheerful connotation than was appropriate for the somber matters with which the project was to deal.

SUNSHINE confirmed that Sr-90 was the critical radionuclide in fallout and recommended that specific measurements of this isotope should be emphasized in future studies. Willard F. Libby of the University of Chicago (well known for his development of methods of archaeological dating by the use of carbon 14, for which he later received the Nobel Prize) attended the meeting and served as leader of many of the discussions.

A satisfactory method for the measurement of Sr-90 in samples of soil, water, or biological material did not yet exist, and Libby, who was a superb radiochemist, agreed to develop a procedure. Both he and J. Laurence Kulp, then at Columbia University, perfected procedures during the next few weeks and opened the way to systematic studies of the ecological behavior of Sr-90. It was agreed that since Sr-90 is similar to calcium in its chemical properties, and will tend to behave like it in soils and biological systems, that it would be convenient to report the results of analyses as microcuries of Sr-90 per gram of calcium (μCi Sr-90/gm Ca). This ratio became known for a while as the Sunshine Unit, but was changed after a while to the strontium unit (SU), which is still used to some extent. Libby later was appointed an AEC commissioner and occupied an important, though often controversial, position as the senior government spokesman on the subject of the public-health implications of nuclear weapons testing in the atmosphere.

During the 1953 conference the estimates of Project GABRIEL were revised on the assumption that Sr-90 would contaminate the calcium pool of the biosphere, and that the dose to the world's population could be calculated from knowledge of the behavior of that element. It was concluded that the limit of "safe" Sr-90 contamination would result from 40,000 megatons of fission, which happened to be the lower limit of GABRIEL.

A big question that emerged from the RAND conference was the whereabouts of the debris from the MIKE explosion. Our worldwide network accounted for only a small fraction of that produced, and our aerial surveys found very little fallout close to the site. We know now that a substantial fraction was probably deposited within a few hundred miles from Eniwetok, but was missed by us because it fell into the vast expanse of ocean between the Pacific islands. Could it be that a major fraction was injected into the stratosphere from which it would take many months or years to fall to the earth's surface? This was what we came to believe, but how could the hypothesis be demonstrated?

One way would be to send instruments into the stratosphere to sample its thin air for radioactive dust. I had long been interested in electrostatic methods of collecting dust samples, and had actually improvised such an air sampler in the kitchen of our first apartment in Philadelphia in 1939. On the return flight from Los Angeles I prepared a schematic diagram of a precipitator that could be lifted by balloon to an altitude of about 100,000 feet, and upon my return I discussed his possibility with LeVine. We knew that the next major series of tests would take place in the Pacific starting in March 1954, and that it would involve the explosion of many multimegaton devices. It would be useful to obtain dust samples from the stratosphere before the March tests (code-named CASTLE) began, but that was only a little more than seven months off. The instrument was only a schematic diagram and it would have to be designed, built, and tested. It would be necessary to arrange for balloon flights, a technology with which we had no experience. Could all of this be done in seven months? It didn't seem possible, but it was done, thanks to hard work by the HASL staff and excellent cooperation from the U.S. Weather Bureau, which arranged for a series of twelve flights to about 100,000 feet. These were completed by late fall. Although too few samples were collected to permit an estimate of the amount of Sr-90 in the stratosphere, we did demonstrate that the nuclide was present. I believe this was the first time dust samples were collected from that altitude but there was no interest by others, although we would have been happy to share our samples so that the characteristics of stratospheric dust could be studied.

Preparations for monitoring the Pacific islands during the tests of March 1954 took much of my time during this period, but it was also necessary to decide the needs of the distant monitoring program. The gummed-film network by then extended around the world, and we were beginning to place occasional stainless steel pots at representative locations so that we could collect total fallout for comparison with the data obtained from the films. As a result of the RAND conference, there began a worldwide search for Sr-90 in all sorts of materials. Samples of rain water, soils, foods, ice cores from Greenland, vintage wines, and old cheeses were only some of the materials that were analyzed to obtain an understanding of the rates of accumulation of Sr-90 in the ecosystems of the world. We quickly came to the conclusion that, because of its similarity to calcium, Sr-90 was present in minute but detectable quantities in all forms of life.

In the fall of 1953 there was little information to guide us in the design of our sampling programs. An important suggestion was made by D. Lyle Alexander of the U.S. Department of Agriculture who noted that the amount of Sr-90 in the milk produced in an area should be inversely proportional to the amount of soil calcium available to the grass consumed by the dairy cows. On the basis of his suggestion, we established a milk-sampling network that included places in which the soil calcium spanned the known levels in the U.S.

Alexander and I were at an agricultural field station at Logan, Utah and decided to take advantage of the fact that some sheep were being confined to a pasture where we could collect samples of soil, manure, milk, and bone. Analysis of the samples for strontium 90 would add to our knowledge of the processes by which the radionuclide passes from the soil to man. We were crawling around the pasture making our collections of grass and manure when he turned to me and asked, "Merril, is what we are doing in your job description?" We were both high up in the civil service system, and I am certain that he was right in his implication that the bureaucrats would have demoted us on the spot if they could have seen us collecting bags of grass and sheep droppings!

The procedure for measuring Sr-90 in minute amounts is very tedious. Although originally developed by Libby at the University of Chicago and Kulp at Columbia, Harley and his associates at HASL deserve much of the credit for perfecting the method, developing a system of quality control, and adapting the method to the routine measurements of thousands of samples.

PREPARATIONS FOR CASTLE

In the fall of 1953 I returned to Pearl Harbor to discuss our participation in CASTLE, which was scheduled to start on March 1 with a multi-megaton explosion from a barge anchored on the reef of what I erroneously thought was to be Eniwetok atoll but actually turned out to be Bikini. I was received warmly by the CINCPAC staff, who I knew were very pleased with our surveys during IVY, in the fall of 1952. We had obviously developed a symbiotic relationship with CINCPAC, even though our interests and objectives were very different. CINCPAC had the practical need of assuring that the inhabitants of the Pacific islands were not affected by the fallout. The HASL objective was to obtain information about fallout because we believed that its effects were being underestimated by both the Department of Defense and the AEC.

The HASL plan was to use the same general methods developed for IVY, except that our ground instrumentation had been improved and would be located on islands on which military weather stations would be maintained during the exercises. These units would be equipped with continuous gamma radiation recorders. The military personnel would monitor the units and report the readings to the HASL coordinator aboard the *Estes*, the flagship for the task force designated JTF-7, again under the command of General Clarkson. The ground-level measurements would assist us in the deployment of the aerial sweeps.

The HASL calculations, based on extrapolations from the fallout encountered following the two JANGLE explosions in November 1951, had indicated that dangerously high levels of fallout could occur on atolls as far away as a hundred miles following the large explosions contemplated for the CASTLE series. I discussed with the CINCPAC staff the need to provide an evacuation capability for the atolls closest to the explosions, but I had not seen the plans for CASTLE and was under the erroneous impression that the tests were to take place at Eniwetok again, as was the case during IVY. This is another example of the problems created by secrecy. I had been sent a list of the contemplated detonations, their expected yields, and the approximate times, but was not given their locations. That was in another document which I didn't know existed and thus did not ask for. One result of this confusion was that at the time of the conferences with CINCPAC, seven months before the start of CASTLE, I thought the nearest atoll to the blasts would be Ujelang because it was the closest to Eniwetok, and I made a recommendation that JTF-7 provide the capability to evacuate the natives of that atoll in the event conditions so warranted. This recommendation was supported by CINCPAC, but the staff of JTF-7 decided that such an evacuation capability would not be needed because the tests would not be conducted if there was any possibility of fallout on the atolls.

At HASL we were faced with the dilemma that if CASTLE was conducted properly, and the meteorological predictions were valid, the fallout would land mainly in the ocean, as apparently happened during IVY. This would be a very desirable result from the standpoint of the natives of the islands, but would not give us the information we needed to demonstrate the danger of fallout in wartime. We conceived the idea of laying oil slicks downwind of the explosions, so that particles that fell on them might be retained long enough to permit them to be measured from the air. This would make it possible to fly over the slicks with our scintillometers and measure the amount of fallout. Experiments designed to test the feasibility of this method were conducted off the New Jersey coast in January 1954, but the ocean was so rough that coherent oil slicks could not be maintained long enough to obtain useful information. We decided to continue the tests in the calmer waters of the Pacific. John Harley, who was already in Hawaii, made plans to obtain the required oil, which would be dropped from low-flying aircraft. We even experimented with methods of dropping the oil. Plastic bags that would rupture on contact? Oil drums that would be dropped and be riddled with bullets? These and other schemes were discussed and some of them tried. In the end we decided that the method simply was not feasible because we couldn't lay an oil slick that would hold fallout particles for more than one hour.

Regrettably or fortunately, depending on the point of view, it proved unnecessary to lay the oil slicks because massive fallout occurred on several atolls about seven hours after BRAVO, the first detonation of the CASTLE series, which was exploded near dawn on March 1.

The preparations for BRAVO were similar to those for MIKE, except that Breslin was assigned to the *Estes,* which permitted me to remain in the laboratory in New York to supervise the oil slick experiments which we planned to continue as long as possible in the hope that it would be possible to use the technique sometime during CASTLE. In addition, automatic monitoring instruments were placed on certain atolls in the care of military personnel. About seven hours post-shot, Breslin received a message from the Air Weather Station on the atoll of Rongrik, about a hundred miles east of the explosion, informing him that our gamma radiation detection instrument had gone off scale, i.e. the radiation level exceeded 100 mr/hr. He immediately forwarded the information to me and attempted to confirm the Rongrik report with surveillance flights by our aircraft based on Kwajalein. However, for reasons that have never been explained, the Task Force prevented Breslin from using the radio transmission facilities aboard the *Estes* for about thirty hours. The operating procedures developed by HASL and approved by JTF-7 called for immediate aerial confirmation but Breslin's hands were tied for many crucial hours after his telegram to me.

At about thirty hours post-shot, the twenty-eight weather station personnel were evacuated by air, but by that time they had accumulated a dose of 78 rem, which is barely below the dose at which the acute effects of radiation would be experienced. More serious was the fact that when the surveillance aircraft flew over other nearby atolls it found that the atoll of Rongelap, about forty miles west of Rongrik, had received even greater fallout. Nearby Utirik and Ailinginae had also received heavy fallout. A decision was made to evacuate the Marshallese natives from the three atolls, and a total of 258 persons were removed by ship, but not until more than fifty hours after the fallout occurred.

When a medical team was sent to examine and treat the evacuated natives, they concluded that the Rongelaps had received whole body doses of about 175 rem, with much higher doses to their thyroids due to the radioactive iodine to which they had been exposed. Some of the Rongelaps have developed thyroid abnormalities, including cancer, as a result of their exposure. If the HASL instrument had not been in place, there would have been no way of knowing that the fallout had occurred, and the doses received by the natives would have been very much higher. On the other hand, had the HASL procedures been followed, the natives could have been evacuated sooner, and the doses received would have been very much less.

The circumstances surrounding the BRAVO fallout have been shrouded in mystery for more than thirty-five years. Published references to the incident frequently refer to it as an "accident" that resulted from an unexpected shift in wind direction. But was the shift greater than normally allowed by the margin of safety provided? Did the forecast properly interpret the available meteorological information? Why was a gag placed on Breslin? Why weren't the confirmatory flights made as required by the written procedures and dictated by common sense? Why was there never a formal inquiry of the episode?

THE LUCKY DRAGON INCIDENT

Unknown to anyone until the fishing vessel returned to its home port of Yaizu in Japan on March 14, the inhabitants of the Marshall Islands and the twenty-eight servicemen on Rongrik were not the only ones affected by the BRAVO fallout. At the time of the explosion, the *Fukuryu Maru (Lucky Dragon) No. 5*, a 100-ton tuna fishing boat with a crew of twenty-three, was about eighty miles east of Bikini when the crew saw a bright flash and realized that they had witnessed a nuclear explosion. They immediately sailed away from Bikini, but after four hours encountered a fallout of white particles so large as to be individually visible. The fishermen later reported that by the time the fallout stopped, the ship appeared to have been coated by a thin layer of snow.

The U.S. had declared the area around Bikini a restricted zone, which was known to the fishermen. The evidence indicates that they were just outside the specified boundary at the time of the explosion and were nervous about being apprehended by the U.S. authorities because the same crew had been found poaching in Indonesian waters one year earlier and had spent some time in jail as a result. It was apparently for this reason that the crew made no mention by radio of the fact that they became sick, starting a few days after the fallout. By the time they reached their home port 14 days later, all the crew were seriously ill.

I learned about the *Fukuryu Maru* the same way most people did, from Japanese news reports that a fishing boat had returned to its home port of Yaizu (south of Tokyo) with twenty-three crew members suffering from radiation sickness. The story did not seem credible to me at first because I was familiar with the extensive precautionary sweeps made over the area in advance of a test. I had in fact flown on one such search and was impressed with what I thought was the thoroughness of the procedure. Nevertheless, Bugher contacted John Morton, newly appointed director of the Atomic Bomb Casualty Commission in Hiroshima, and asked him to obtain whatever information was available. I knew also that John Harley was in Japan, and attempted to alert him to the fact that he might be needed but he had already departed for home.

It took very little time for Morton to confirm from his Japanese colleagues that the men were suffering from radiation sickness, and that the boat was highly radioactive as was its cargo of tuna fish, which had been disposed of by burial. On the morning of March 19, Bugher and I participated in a conference call from Morton who said a consultant was badly needed to advise the U.S. embassy as well as the Japanese on the radiological aspects of the matter. It was decided that I should proceed to Tokyo immediately.

That call came at a particularly awkward time. I was attending an important meeting of the NYOO staff at which it had just been announced that the office would be reorganized and that Burke Fry, manager of the NYOO, was to retire. Alphonso Tammaro, the Assistant General Manager for Research and Development on the Washington staff, was present at the meeting, and when it became apparent that I had to leave, he and Fry drew me aside to inform me that I had been selected as the new Manager of Operations. That came as a total surprise. I responded that while I was flattered to be asked to assume greater responsibility, I thought it important that I remain as head of HASL, at least for a while. They suggested that I could fill both positions if I wished to do so. I agreed, quickly took off, and didn't have a chance to think about the matter in the weeks ahead. That was one of the major decisions in my life, but they proposed and I accepted in no more than five minutes!

In an organization like the U.S. government it is amazing how fast things can move when an emergency arises. For security reasons, the passports of all AEC employees in those days were kept in the Washington headquarters, and preparation for foreign travel normally required about one month during which the travel application form crept from one desk to the next collecting authorizing signatures. The call from Morton came at 11:30 A.M., at the conclusion of which it was decided that I should proceed to Washington for a conference at AEC headquarters before leaving for Japan. It took me a few hours to collect the instruments and reference material I knew I would be needing, and I left on the four P.M. plane from La Guardia where Irma, my mother, and our three boys were waiting for me with the bags she had packed. When I arrived in Bugher's office, he had my passport, my authorization for travel to Japan, my tickets, and an ample travel advance. We discussed the important implications of what had happened in the Pacific, and he told me that about one million pounds of tuna were suspected of being contaminated. This added a new dimension to an already complicated situation.

A flight to Tokyo from the East Coast now takes about fourteen hours, and can be nonstop. In 1954, it took twelve hours to reach the West Coast, about ten more to Honolulu, another six or eight hours to Wake Island, and a final ten-hour leg to Tokyo. Forty hours total, with an additional three or four hours consumed at the stopovers. But travel was much more comfortable, and the planes even had sleeping berths for those that wanted them. We reached Honolulu at 5:30 A.M. and only then did I find that the *Lucky Dragon* affair had blossomed into a full blown international incident. I was met at the airport by Commander Deller, who was on the CINCPAC staff and was my liaison with JTF-7. He briefed me on the latest developments in the Marshalls. A medical team from the U.S. had arrived there and were keeping the evacuated natives under observation. Rumors about the evacuation had reached the international press, abetted of course, by the sensational stories from Japan. The problem was exacerbated, so far as I was concerned, by the announcement of a small pharmaceutical company that they were sending a drug to Japan that would help the fishermen. The medicine was no more than a common over-the-counter skin burn lotion, but Bugher, who of course knew nothing about the shipment, had cabled prominent members of the Japanese medical profession that I was on my way to Japan to assist them, and when they learned of the announcement of the publicity-seeking pharmaceutical company, it was assumed that I had some miracle drug with me. As I left Honolulu, I was surprised to read about this in the *Honolulu Times*, and it disturbed me more than a little.

I reached Tokyo at 10:30 P.M. totally unprepared for the reception I received. After the plane had rolled to a stop, one of the stewardesses tapped me on the shoulder and introduced me to an officer of the U.S. Military Police, who was my escort for the next few minutes. He asked for my passport and baggage checks, and told me that there were representatives from the press waiting for me, but that I should make no statements. Because I had been sitting on the opposite side of the plane, I was not aware of the large number of reporters and photographers waiting for the plane and, had I seen them, I would have assumed they were waiting for another passenger. There were about thirty of them, behind a barricade at the edge of the apron. They were close enough to take pictures, which appeared in the newspapers the next morning, but except for the few seconds it took for me to descend from the plane, they saw no more of me that evening. An embassy sedan was at the bottom of the ramp and I stepped into the rear seat where I found William Leonhart, who introduced himself as First Secretary of the U.S. embassy. Without waiting for my baggage, the car took off for the Sanno Hotel, at which I had stayed during previous visits. It had been taken over by the United States Army during the occupation of Japan to billet officers in transit.

En route, Leonhart briefed me efficiently about the state of things. The Japanese people were angry about the fallout on the *Lucky Dragon*. They were the only people ever to have been hurt by atomic bombs, first during the war, and now again. The Japanese scientists were making sensational statements to the media. The formal peace treaty between the U.S. and Japan had only recently been signed, and this was the first postwar crisis in the relations between the two countries. The bottom had dropped out of the tuna market. No one in Japan fully understood the technical implications of the event, and I should expect to spend considerable time providing information to both the Japanese and the Americans.

After a restless night I had a breakfast conference with John Morton, who was also staying at the Sanno. He was in a very difficult position. His entire career had been spent as a surgeon on the staff of the University of Rochester Medical School, from which he had recently retired as chairman of the Department of Surgery. He knew nothing about radiation medicine, but was invited to come to Japan because he had developed a good reputation as a scientific administrator. He was in no position to answer the kinds of questions that were being asked. How should the doses to the fishermen be calculated? What radioactive substances were in the fallout? What was the allowable level of contamination in tuna fish? How is radiation illness treated?

After breakfast we began a round of conferences and visitations that continued for three weeks. I spent the first hour reading the cables that had been going back and forth between Tokyo and Washington, from which I could sense the deterioration in the relationship between the two countries during the few days in which Japanese scientists, newspaper writers, and some officials had vented their fears and hatreds against the U.S. and its military forces in Japan. These troops had been an army of occupation from 1945 until the formal peace agreement was signed in 1952, but the U.S. still maintained a large military establishment in Japan. This was the first serious interruption in the otherwise smooth relationship that had existed between the two countries since the end of World War II. It was suggested by some Americans that the fishermen were spying on the U.S. bomb tests. Could this have been so? In the end there was no evidence to support the allegation. Were the fishermen as sick as claimed by some of the Japanese physicians? They were indeed sick and getting sicker by the day. Could the U.S. provide medical assistance? Very little. After all, the Japanese had been through the atomic bombings of two cities only nine years before, and many of the physicians involved in treatment of the survivors were now treating the fisherman.

However, the Japanese had no understanding of the biophysics of radioactive contamination, and they needed and wanted assistance in that area and welcomed any help I could give them. Immediately after my arrival, the Japanese scientists expressed a desire to meet with me. Unfortunately, there was a considerable amount of rivalry between different groups of scientists. The staff at Tokyo University was at odds with the group at the National Institute of Health, and the local physicians in Yaizu were unhappy because some of the fishermen had been transferred to a hospital in Tokyo. The Japanese government had appointed an official committee to investigate the incident and recommend the steps that should be taken. The ambassador was advised that all communications of a scientific nature between the U.S. and Japanese should be through that committee, which was chaired by Dr. Rokuzo Kobayashi, Director of the Japanese National Institute of Health. I met with his committee on March 24, accompanied by Morton, and representatives of the Far East Command and the embassy. It was a difficult conference because very few Japanese scientists spoke English and the interpreters from the diplomatic offices were not familiar with many of the scientific terms being used.

It was at that meeting that I had my first example of the misunder-standings that can arise from subtle errors in translation. Dr. Maseo Tsuzuki, a physician who had been barred from his position as professor of surgery by General MacArthur because he had held the rank of rear admiral in the Imperial Navy (i.e., he had been "purged" in the vernacu-lar of the times), was in Yaizu on the day of our meeting, but left word with one of the committee members that he would return to his home by eleven o'clock that evening and that it was urgent that I call on him. That seemed a strange time for a visit, but since the message was transmitted by a member of the Japanese Foreign Office, arrangements were made for an embassy car to pick me up at the Sanno in time for our meeting in his home. When we arrived the house was dark, but our knock on the door wakened Tsuzuki, who received us in his kimono. It turned out that he did want to speak to me, and had left a message that I should *call* him at the designated time, not *call on* him. Fortunately he spoke enough En-glish to understand the humor of the situation and we sat for more than one hour getting acquainted. That private conversation early in my visit was very fortunate because Tsuzuki became very relaxed as we sipped warm saki. I developed an understanding of him that was to prove useful in the difficult days ahead, and would lead to a close friendship that lasted until his death in 1961.

Tsuzuki had been one of the most controversial of the Japanese scien-tists and had not shown a willingness to accept the assistance of the ABCC staff, represented by the director, John Morton. Tsuzuki was sur-prisingly frank in telling me of his resentment because the U.S. occupa-tion forces had confiscated a report he had written following his survey of the effects of the bombings in Hiroshima and Nagasaki. He had led the team of Japanese physicians that entered the two cities to provide medi-cal assistance, but he was never allowed to publish his report. I was fa-miliar with that report, which actually had been translated into English and published, under his authorship, as an appendix to a report issued, but not widely circulated, by the U.S. National Research Council. It was true that Tsuzuki's report, which was a classic, did not receive the recog-nition that it deserved.

During that hour-long conversation, Tsuzuki also discussed his unhap-piness about the fact that he had been placed on MacArthur's list of for-mer officers that were purged from their civilian positions. I reasoned that in the aftermath of a long and cruel war there were bound to be inequities, but I knew that he was respected as a medical scientist in the U.S. and promised that I would look into his status. (This I did, and with such success that he was invited to the U.S. by the State Department a few months later.) In addition I assured him that he would be free to publish his studies of the injured fisherman, and that my position was that of a U.S. scientist and government official who was there to advise my government about the circumstances of the accident and assist the Japanese scientists in any way possible.

Following the long day of conferences, first with the committee and then with Tsuzuki, it was clear to me that there were a number of separate but interrelated problems to be addressed:

1. The clinical management of the twenty-three fishermen: In my opinion there was little help to be offered. The U.S. had no methods of treating acute radiation sickness that were not already known to the Japanese physicians.
2. The dose received by the fishermen: This included that delivered by external radiation because they lived for fourteen days on a ship covered with radioactive dust, and by internal radiation because the fallout particles were inhaled or ingested.
3. The concerns of the Japanese that the Pacific tuna would be contaminated.

By the time of my arrival, the Japanese physicists had already estimated the external radiation dose using crude instruments that were nevertheless quite reliable in their expert hands. However, they did not know how to estimate the dose delivered by radionuclides deposited in the bodies of the fishermen. In 1954, even the radioactive species present in bomb fallout were secret, but by the time I arrived, Japanese radiochemists had made progress in analyzing particles of fallout collected from the *Lucky Dragon*. However, their results were purely qualitative, and they were unable to separate nuclides that had similar chemical properties. They were fortunate to have the assistance of Professor Kenjiro Kimura, an internationally respected radiochemist who had attracted attention after the bombings by concluding, correctly, that the Nagasaki bomb utilized plutonium because he had found traces of that element in samples of soil collected from an area in which rainout had occurred. He had also discovered previously that he could produce U-237 in the laboratory by bombardment of U-238 with fast neutrons. In my first meeting with him he told me he had found U-237 in the "Bikini ashes" from which he concluded that the March 1 explosion involved the fission of U-238, which was still a secret known by only a few scientists back home.

In response to my cable to him, Harley advised that the isotopic content of the bomb debris would be similar to the tables of fission products that had been published in the open literature, from which the Japanese scientists concluded that Sr-90 was the isotope that would be of greatest danger to the fishermen.

I had requested samples of urine from the fishermen with the understanding that I would send them to Harley for analyses at HASL. The amounts of the various nuclides in the urine would give us a clue as to the quantities that were deposited in their bodies. On the basis of the results received from the laboratory I advised the committee that the amount of radioactivity deposited in the bodies of the fishermen was insignificant in relation to the dose received externally.[4] This came as a big surprise to me because the men had lived for fourteen days in an environment contaminated with radioactivity to an unprecedented extent. In fact, the analytical results were so unexpected that I initially wondered about their validity but any doubts were dispelled a few months later when one of the fishermen died of serum hepatitis and his tissues were analyzed for the major radionuclides by Professor Kimura. He confirmed our findings: only insignificant amounts of radioactivity were found in the body of the deceased fisherman.

Shortly after my arrival in Japan, General Hull, who had replaced Douglas McArthur as Supreme Commander of the Allied Forces in the Pacific, asked to meet me and some of his staff at their headquarters in the Dai Ichi building. By the time of our conference, the true importance of the BRAVO fallout seemed obvious to me: Thermonuclear weapons had the ability to contaminate tens of thousands of square miles with lethal amounts of radioactivity. This was what we at HASL had suspected but now we had the proof.

In my briefing of Hull I made it clear that I was in no position to discuss the full military and political implications of the Bikini fallout, but it was quite obvious that thermonuclear weapons were far more destructive than had been anticipated by the military planners with whom I had been discussing this possibility for the past two years. Hull didn't pursue this point, which was understandable, considering my position. However, he did ask repeatedly about the security implications of the fallout residues present on the ship. He was aware that a radiochemical analysis of the particles of fallout could reveal classified information about the design of the BRAVO device. If it was in the U.S. interest to do so, he was prepared to seize the fishing boat. I was surprised to hear him speak so strongly about the security aspect because it was obvious that most of the fallout had already been removed from the boat. Much of the material had undoubtedly been passed to innumerable scientists throughout Japan and possibly elsewhere. As a matter of fact, I had in my pocket a vial of fallout particles that had been given to me by one of the Japanese scientists that very morning. I told Hull that the people who were in the best position to advise on the security implications were at that very time with JTF-7 at Eniwetok, and that I would consult with them as to whether any action was required to secure whatever fallout particles remained on the boat. I sent a radiogram to General Clarkson asking for the opinion of the Task Force intelligence personnel, and when he replied promptly that the boat was of no interest, the matter was put to rest.

During the first week of my visit, much of the time of the embassy staff, as well as Morton and myself, was spent negotiating with the committee to get permission for American physicians to examine the fishermen. The Japanese scientists were reporting daily on their medical status. The condition of the men was said to be deteriorating: they had lost their hair, had developed skin ulcers from burns caused by the beta particles emitted from the fallout that deposited on their skin and, more ominously, their white blood counts were continuing to decline. The interest of the ABCC in having contact with the cases was understandable. ABCC was responsible for one of the largest medical follow-up studies in history, which was concerned with the delayed effects of exposure to radiation. The twenty-three Japanese fishermen were the largest group of persons suffering from the immediate effects of radiation since the bombings of World War II. In the United States, physicians who were interested in developing methods of treating acute radiation injury wanted to collaborate with the Japanese physicians. But the Japanese were adamant that they wanted no help in dealing with the patients, although they did welcome any assistance I could provide that would help them to understand the physical and radiochemical problems they were facing.

I was anxious to visit the *Lucky Dragon,* and the committee not only arranged for this, but suggested that while we were in Yaizu we should pay a courtesy call on the twenty-one fishermen who were hospitalized there. On March 26, Morton, Dr. Lewis, who was an ABCC hematologist, and I, accompanied by several Japanese physicians and physicists, flew to Yaizu, ninety minutes southwest of Tokyo, in a C-47 provided by the U.S. Air Force.

When we arrived, we were welcomed impassively by the school children of Yaizu, who were lined up along the short airstrip. We proceeded to the hospital followed by a few carloads of reporters who had also met us at the field. The visit to the patients was only a courtesy call; they were evidently glad to see us. They were resting on mats, surrounded by their families who, in the Japanese custom, were preparing food for them on hibachis within the room. The reporters were not allowed into the hospital, but since the windows were at ground level and wide open, the photographers had no trouble taking all the photographs they needed. I had broken the Geiger-Mueller tube of one of the two radiation detectors I had brought with me, but with the remaining instrument I was permitted to scan the bodies of some of the fisherman. Although it was now nearly four weeks since the accident, their thyroids still contained readily measurable amounts of iodine 131. The information I obtained was very scanty because there was not time for more systematic measurements. Many of the men had skin burns, particularly on their scalps and along the line of their trouser belts where the fallout particles had become lodged as they stood shirtless on deck when the fallout was occurring.

For some reason I never understood, the American media carried reports that I was not permitted to examine the fishermen because I had neither an M.D. or Ph.D. The fact was that the Japanese scientists requested me to visit the fishermen when we visited Yaizu. Unreliable media reports can be very troublesome.

We spent no more than thirty minutes with the fishermen and then went to a picturesque bay-front hotel where we were greeted by the mayor of Yaizu who had arranged what must have been a very expensive lunch for the mayor of a small fishing community in postwar Japan.

After our lunch we visited the fishing boat, where the Japanese scientists who accompanied us donned lab coats and gauze masks before boarding. We Americans brought no protection, nor did I believe it was necessary for the type of contamination we were to encounter. However, when we saw our pictures in the Tokyo papers next morning, I did regret that, compared to the Japanese, we seemed so cavalier in our attitude towards radioactivity. Before leaving Tokyo I had requested that the embassy provide me with a household vacuum cleaner, and this I used to collect fallout particles from some of the less accessible exterior surfaces of the boat that I thought might have escaped the pickings of the many Japanese scientists who preceded me. On the roof of the cabin I found a loose piece of wood about eighteen inches long that was coated with many white grains about 0.2 mm. in diameter. With the permission of the Japanese, I took the board back to HASL as a souvenir of my visit. The dust collected by the vacuum cleaner was divided for study by several laboratories in Japan and the U.S.

The first opportunity to meet members of the committee socially came after about one week, when William Leonhart and his wife entertained the Japanese scientists at dinner. He and I were busy at the embassy and didn't arrive at his home until exactly 6:30 P.M., when the reception and dinner were about to begin. We found that most of the Japanese were already there, my first experience with the customary promptness of the Japanese people. It was a pleasant evening but the language problem was insurmountable. Thirty years later I was entertained at a reception in Tokyo following a talk in which I was invited to reminisce about the events of that period. Several members of the committee attended that pleasant reunion and I was greatly impressed with the ease with which we communicated in English.

During the last week in March, nearly one month after their exposure to the fallout radiation, the white blood counts of most of the fishermen were continuing to decrease, and alarm increased over the possibility that some of them would not survive. In one case the count dropped to 800 cells per cubic millimeter and others hovered just above 1000. This is a dangerous phase of the acute radiation syndrome, in which the body is less able to resist infection. If infection can be prevented for a few weeks, the damaged bone marrow, which produces a reduced number of blood cells, will gradually repair so that recuperation of the white blood count can begin. The ABCC physicians were of the opinion that they could possibly make practical suggestions that would reduce the probability of infections, but the Japanese doctors continued to decline their assistance, and neither Morton nor his American assistants were ever consulted concerning medical management of the cases. Fortunately, the blood counts began to return to normal after four to six weeks, and the men returned to normal health after one year, except for the radio operator who developed jaundice in June (about three months after exposure) and died in

September. The cause of death has been attributed to serum hepatitis, probably a consequence of the large number of blood transfusions he received. It was accepted practice in Japan at that time to transfuse only about 100 cc at a time. This required a great number of transfusions, with a proportionate increase in the risk of infection by the hepatitis virus. Although the radio operator did not die directly from radiation injury, his death was clearly a secondary result of his exposure.

The Tuna Panic

The *Lucky Dragon* incident had implications beyond concern over the health of the crew. The *Lucky Dragon* had landed with 28,000 pounds of fish that were quickly disposed of by burial. When reports were received of the *Lucky Dragon* incident, the U.S. Food and Drug Administration decided to monitor incoming shipments of tuna, which was a sensible precautionary step and would have created little problem because there was no general contamination of tuna. However, the U.S. tuna companies sent notice to Japan that they would not pay for shipments of fish unless they were certified as "nonradioactive" before the shipments left Japan. Consumption of tuna in Japan dropped immediately as a result of this action. Concern—unwarranted, as it turned out—about the radioactivity of tuna had serious economic consequences in Japan, where the tuna fleet consisted of about 1000 vessels, with an annual catch estimated to be worth twenty-six million dollars.

When I arrived on March 22, the Japanese had already monitored and cleared the first outgoing shipment of frozen tuna. Technicians had been trained in the use of Geiger counters, and were assigned to the five major ports at which tuna was received and shipped. All fishing boats were instructed to deliver their catches to one of the designated ports. Radiation detection instruments were loaned to the Japanese by the U.S. Far East Command.

At a conference with Japanese officials two days after my arrival, they requested that I recommend monitoring procedures and standards to determine if the tuna could be cleared for shipment. Because of my unfamiliarity with the manner in which fish are processed in the course of shipment, I requested and was given permission to inspect the loading of tuna at a dock in Yokohama. I was flabbergasted by what I saw. Hundreds of tons of frozen tuna were moved from a refrigerated warehouse to the dock, from which they were promptly transferred to the hold of the ship. With only a few instruments available, and in the hands of inexperienced inspectors, how could the huge cargos be surveyed?

I suggested that, as a first step, measurements be made of one in every ten fish before they were loaded on the ship. The fish should be examined for one minute by passing a Geiger counter over its surface, with particular attention to the gills, because they filter large volumes of water during the process of respiration. The instrument probe should then be inserted into the mouth of the fish, and into the abdominal incision through which the viscera had been removed. Additional instruments would be needed and the number of trained inspectors would need to be increased.

There remained the question of the criteria for rejection of fish as contaminated. It is not a simple matter to estimate the risk to consumers of fish from measurements made in this way, and I told the Japanese officials that I could not recommend a standard without further study. However I did not believe fish with more than insignificant contamination would be found. Low levels of radioactivity on the skins and gills were a possibility, but this would not be important to the canners, who routinely strip the skins and remove the heads as the first step in processing. I suggested to the Japanese that since I would be in Tokyo for several more days, I wanted to be informed when contaminated tuna were found. I would arrange for immediate air transport as needed to any of the five ports that had been designated to receive and ship tuna to the states. My recommendations for certifying the tuna would depend on what I found.

Many reports of contaminated tuna appeared in the newspapers, and several times I was alerted that I should fly to one or another of the five ports, but each time I was soon informed that we had received a false alarm and that no trip would be necessary. Between March 24, when the Japanese Foreign Office first discussed the matter with me, and April 9, when I left Japan, there were many such reports but I never saw radioactive fish. In some cases the reports were received after the catch was dumped at sea. It was suggested to me that the refrigeration systems on the boats sometimes failed at sea, causing the fish to spoil. With all the publicity being given to radioactive fish, a resourceful captain could dump his cargo and later claim that he did so because it was radioactive! This would make it possible for him to prepare a claim against the U.S. government. The suggestion was only anecdotal and I have no way of knowing whether it explains why I never saw radioactive fish despite the fact that I had a C-47 and helicopter at my disposal and could quickly reach any of the five Japanese tuna ports.

WHATEVER WAS HAPPENING IN WASHINGTON?

During my stay in Japan I had excellent rapport with the AEC in Washington as well as with HASL in New York. There was hardly a day when I didn't cable for information, copies of literature requested by the Japanese scientists or, in some cases, equipment. I had excellent support and all my requests were answered promptly. My only problem was personal, because Tokyo was ten hours behind the East Coast of the U.S. Soon after I tumbled into bed at 11 P.M. after a long hard day, my colleagues were reporting for work in Washington and New York with a thirst for the latest information from Japan. Each day I helped the embassy staff prepare a detailed telegram to Secretary of State Dulles. This was dispatched in the early evening Tokyo time, but could not possibly provide all the information that was needed, so many of the calls I received during the night were unavoidable. Had I anticipated the demands for information, I would have asked for assistance in Japan by other radiological specialists from the States, but by the time I realized how weary I was getting, it was too late to obtain help. The ambassador recognized what was happening and arranged to screen the night calls so that I could get some rest. One day he insisted that I get away for twenty-four hours and arranged for me to be driven to a hotel near Mount Fuji which the U.S. was using as a rest and recreation center for field-grade officers. The hotel cost me one dollar for a twenty-four hour stay with meals. Thirty years later, when it had long since been returned to private hands, I stopped with my wife at the same hotel and the rate was $100 per person per night!

My greatest concern with Washington was the near absence of official statements. A terse statement that a new test series had begun in the Pacific was issued to the press immediately following the BRAVO detonation, but no other information was given at that time despite the fact that it was already known that unexpectedly high levels of fallout had occurred. No further information was released for many days. The evacuation of the natives and the dispatch of a team of medical specialists all took place without public announcement. But JTF-7 was a task force of many thousands of personnel, and the magnitude of the event was such that it was impossible to keep rumors of the disaster from reaching the public. It was not until March 10 that the AEC issued its second report in which it announced that 236 Marshallese had been evacuated from their home atolls "according to plan as a precautionary measure."[5] This was a clear understatement of the facts. The statement said nothing about the early effects of the heavy radiation exposures that were already evident.

The magnitude of the fallout became known everywhere in the world on March 16 when the Japanese newspapers announced that the *Lucky Dragon* had returned to its home port of Yaizu with twenty-three fishermen suffering from radiation sickness. The news created a sensation throughout the world, but no immediate statement was issued by Washington. In the absence of information from the U.S., newspapers in western countries, including the United States, were forced to rely on reports from Japan which were in many cases exaggerated.

On March 24, shortly after my arrival in Tokyo, the U.S. government announced that it was expanding the restricted area around the Pacific test site, and that the surveillance procedures were being increased to provide greater assurance against unauthorized entry into the restricted zone.[6] The statement went on to say that because of the slow movement of the water around the Marshall Islands, the radioactivity would become harmless beyond a few miles, and would be undetectable within 500 miles or less. This last statement bothered me greatly because while I did not believe the contamination levels in the Pacific Ocean would be hazardous to consumers of tuna fish, the radioactivity would certainly be detectable.

By March 31 there were no additional announcements from Washington and Ambassador Allison decided it would be necessary for me to conduct a background briefing for the American members of the Tokyo press corps. Since it was assumed that the reporters, although a sophisticated group, would know little or nothing about fallout, contamination of ecosystems, the effects of radiation exposure and, in particular, how the fallout occurred, I was asked to provide a two-hour tutorial as well as the opportunity to answer questions they would ask. It was understood that I could not divulge classified information, but I did not think this was a handicap for them or myself, since secret information about the BRAVO device itself would add nothing to their understanding about the circumstances of the fallout. Besides, I knew next to nothing about the classified aspects of the matter. The briefing was off the record, which meant that the reporters could use the information to interpret what they were hearing, but could not quote what they were being told and could not refer to the fact that they had attended a briefing arranged by the embassy.

What I could not explain to the reporters, because I did not know, was why our government had released so little information. It was now a full month since the fallout occurred, and about two weeks since the *Lucky Dragon* had returned to Yaizu. The Japanese scientists were making statements that were being reported in the world press. The statements they were making were frequently inflammatory, which was understandable under the circumstances. Sensational stories were being published in the U.S. newspapers as well. The ambassador and his staff were puzzled that so little information had been released.

The press briefing took place during the afternoon of March 31, Tokyo time, which was early on March 30 in Washington. Unknown to me, President Eisenhower had scheduled a televised press conference for the following day and had invited Admiral Lewis L. Strauss, chairman of the Atomic Energy Commission, to join him and to make a statement about the BRAVO fallout. I believe that it was a coincidence that the briefing in Tokyo coincided with Strauss's statement. The president was responding to the same pressure for information that caused the ambassador to arrange the briefing. When the Strauss statement, which was made in the presence of the president, was received in Tokyo, it came as a shock to Morton and myself as well as to senior officials at the embassy. The statement said the fishermen "must have been well within the danger area," which could not be supported by any evidence of which I was aware, but was a convenient assumption from the point of view of the liability of the U.S. for damages sustained by the fishermen and the owner of the *Lucky Dragon*.[7]

The statement then went on to say that the skin lesions "are believed to be due to the chemical activity of the converted material in the coral, rather than to radioactivity.... " In other words, the burns were due to the fallout of corrosive particles of calcium oxide produced by the action of the great heat of the fireball on coral. This was particularly painful to read since it was not so, and because in the same paragraph that contained that statement, Strauss mentioned that the commission was represented in Japan by Morton and myself. There was every reason why the Japanese scientists, whose confidence we were struggling to develop, should have believed that the misinformation originated with us.

On April 9 Morton and I decided that our usefulness was at an end in Japan and arranged to leave for what I thought would be a brief visit to Eniwetok to meet our colleagues who were ministering to the needs of the Marshallese. The ambassador issued a press release in which he reviewed what we had accomplished, and the foreign minister delivered a note of thanks to us together with gifts of appreciation. I was concerned that a large cloisonnè vase he presented to me would be too heavy to carry home and tried to give it to one of the secretaries at the embassy who declined it with the advice that it was a beautiful piece of art that my wife would certainly enjoy. This turned out to be so.

One reason why I wanted to leave it behind was that I had promised our three sons that I would bring home some rice birds (Java sparrows), attractive gray and pink birds that can be easily tamed. I had found a dealer who had some week-old birds and bought a small bamboo cage containing six fledglings that required feeding every two or three hours during the day. I left Japan with both the large vase and the cage of birds, having arranged with Pan Am to keep the cage in the cabin with me.

The story of the birds is in itself a saga. I thought I would be in the Marshall Islands for only a few days but it turned out that I remained for one month during which the birds outgrew their cage twice. All six of the birds eventually arrived home with me, by which time they were full grown and brought many months of pleasure to our family. The birds were all given Japanese names, often had the run of the house, and frequently flew away on brief excursions that usually ended when a neighbor called to say that one or more of them had interrupted a picnic by lighting on a guest's shoulder!

The fallout of March 1 had proved that we were correct in our belief that massive fallout covering large areas of land would be a consequence of nuclear war. But in my absence from the laboratory, the staff recognized that it was important to obtain more quantitative information. It was necessary to define the area of fallout and the fraction of the radioactive debris that would be deposited. Our experiences had taught us that a nuclear explosion creates clouds of debris that behave in different ways. If a bomb explodes so high above the ground that the fireball does not touch the ground, the particles formed are so small that they deposit very slowly and do not create a serious danger. If the bomb is big enough (greater than 100 kilotons of TNT equivalent), much of the debris will penetrate into the stratosphere, from which it will deposit gradually over a period of many months, by which time the short-lived radionuclides will have decayed. The fraction that does not reach the stratosphere will remain in the lower atmosphere from which the particles will deposit over a period of a few weeks, largely with rain or snow. Such fallout will contain some of the short-lived radionuclides, such as radioactive iodine, but the dose to people will not be sufficient to result in radiation sickness. However, if the fireball touches the ground, or if it is exploded so close to the ground that relatively large particles of smoke or soil are convected into the mass of gases as it cools, the radionuclides will deposit on the surface of large particles which will deposit within a few hours. It was important that this fraction of "close-in" fallout be better understood so that the dangers of nuclear war could be better understood.

William Harris and Harry LeVine, with typical HASL ingenuity, had reviewed the results of our failed oil slick experiments and had decided for the remainder of the CASTLE series to substitute styrofoam rafts that could be dropped in the expected fallout area, and could be surveyed from the air. They had calculated that a floating four foot square slab of styrofoam would be sufficient, and in the excitement of the weeks following BRAVO had no trouble in convincing the Air Force to supply a pair of giant C-97s to transport hundreds of the rafts of Kwajalein. Each raft was equipped with a radio transmitter that would broadcast a homing signal to assist the survey aircraft in finding it. My staff didn't bother to tell me about the plan, called Operation DUMBO, until I arrived in the Marshalls. The plan was well conceived but there was simply not sufficient time to develop all the details. After three weeks of frustrating trials, the project was abandoned.

In between tests of the feasibility of DUMBO, Morton and I caught up with the status of the Marshallese who had been exposed to fallout and, in turn, briefed our colleagues on the condition of the fishermen. Even on so somber an occasion there was time for fun and good humor. On the evening of our arrival a barbecue was arranged on an Eniwetok beach, and Morton and I were inducted into the "Eniwetok County Medical Society." He, being a physician, was made a full member and I, the non-physician, was inducted as an associate member. This was not my last visit to the Marshalls, but it certainly was my most memorable one. I had several days to visit a number of atolls on which HASL had installed radiation detection equipment. I was able to see first hand how the natives lived, played, and worked on their remote strings of islands that circled cobalt blue lagoons like necklaces. Each of the palm-covered islands rested on circular coral reefs that could be seen easily from the air through clear turquoise water. It was particularly interesting to hover over the reefs in a helicopter from which could be seen the intricate surge channels carved through the coral by the pounding surf.

The most thrilling parts of my visits to the atolls were always the underwater explorations of the reefs. Scuba diving techniques were just coming into vogue and the lagoons were the perfect places for being introduced into the marvelous world of living coral and the dense populations of interrelated colorful creatures it supports. The water is so clear that one can see for forty or fifty feet as though looking through unpolluted air, although sometimes the diver is enveloped by great schools of multicolored fish, and when this happens visibility can drop to zero for a few seconds. The giant clams, waving purple mantles, intricate coral shapes, giant rays, and countless brightly colored fish all combined to produce an unreal sensation at being in another world. I have since had opportunities for underwater exploration in many other parts of the world, but I have never seen anything to match the beauty of those undisturbed Marshall Islands coral reefs, in water that must surely be as clear as any in the world.

During those few days of Eniwetok I had time to reflect on what had happened in the previous weeks. I speculated on why the AEC had taken so long to announce what had happened, and why the press releases were so misleading. I have never learned what happened, but I can only assume the AEC was unable to obtain the required agreements of other agencies such as the Departments of Defense and State. The fallout episode must have caused consternation at the Defense Department because they learned that it would be necessary to shelve the plans which, in the event of war with the Soviets, called for destruction of Eastern European air bases with megaton bombs that would create giant craters when exploded at ground level. They were building bombs they couldn't use! That is, unless they wanted to cover western Europe with lethal levels of fallout! The North Atlantic Treaty Alliance (NATO) had only recently been formed and was depending on the U.S. nuclear shield to defend Western Europe. What would be the effect on the NATO alliance of the knowledge that thermonuclear bombs could produce such heavy levels of fallout?

I also reviewed the experiences I had had with the nuclear weapons testing program during the previous three years. The sophistication of our fallout studies program had increased from analyzing hastily collected fresh snow in the Northeastern states after the report we received from the University of Rochester, to measurement of the worldwide patterns of fallout, using sophisticated instrumentation of our own design and advanced radiochemical methods of analysis. We had been proved correct in our belief that the explosion of megaton bombs could cover thousands of square miles of territory with lethal levels of radioactivity.

Because the instrument we placed on Rogerik provided prompt notice that heavy fallout had occurred after BRAVO, the doses received by the American airmen and Marshallese natives were much less than would have been the case had we not alerted the Task Force to the fallout. Some of their lives had undoubtedly been saved, despite the delay in implementing emergency procedures.

Nuclear weapons were abhorrent to me, as they still are, but they existed and it was important that their effects be understood. Our studies were adding to that understanding, and perhaps the prospect that extensive areas of land would be blanketed with radioactive dust would be one more reason why nuclear weapons would never be used in war.

In early May 1954, after an absence of nine weeks, I returned to New York to assume my new responsibilities as manager of the AEC's New York Operations Office. I had been so busy that I had had little time to think about the day on which I had received the phone call from Tokyo, just as I was being offered the new position. I looked forward to it with mixed feelings. At the age of thirty-nine I was being promoted into one of the more important positions within the AEC. I would be involved in the administration of large-scale government research and development contracts and would have the opportunity to participate in the application of nuclear energy to civilian needs such as the production of energy and exploration of outer space. The immediate impact of this change in my career would be softened, however, by the fact that I would continue to serve as the director of HASL.

SIX

Atomic Energy Commission

From Science to Administration

I RETURNED TO New York in early May of 1954 to find NYOO badly demoralized by the reorganization that was underway. When I left in mid-March, the staff had numbered more than 450 (excluding about 75 in HASL) but was to be reduced to 125 by July 1955 by the transfer of responsibility for uranium production functions to the Oak Ridge operations office. Many of the staff moved reluctantly to Oak Ridge, some elected to retire, and some were finding other positions in the New York area. About one month after my return from the Pacific, Burke Fry left the post of manager and I succeeded him. Joseph C. Clarke, a career government administrator with a background in accounting, was appointed as my deputy, and we began a successful and efficient relationship that lasted until 1959, at which time I retired from government service and he succeeded me as manager. With Joe Clarke as my deputy, I found it possible to serve as both laboratory director and manager for the next three years. It was an interesting period in which I was HASL director and also the regional administrator for the Atomic Energy Commission.

I had spent all of my career until then in expanding organizations, and this was my first experience with a major cutback in staff. Beginning on the day of my return, I was consulted daily by staff members who hoped I could ease the disruption in their lives, but in most cases there was little I could do. However, as often happens in government, the reduction in staff was not so traumatic. NYOO ended up very close to its original strength because new responsibilities were gradually added at about the same rate as other activities were being transferred to Oak Ridge.

I had accepted the position of manager of the New York Operations

Office with mixed feelings. I knew it would mean an advancement within the AEC structure, and provide me with a broader base of experience, but my heart was really in the laboratory. I was fortunate that Clarke was capable of relieving me of many of the routine time-consuming managerial duties. This was no small responsibility: At the end of the NYOO reorganization we were administering 275 contracts, of which ten involved the construction of major new research facilities.

Shortly after I began to serve as both manager of operations and laboratory directory, it became apparent that I needed a deputy laboratory director who could assume the directorship if required to do so. This presented a serious problem. The logical person was Bill Harris, who had a background similar to mine and had served as my principal assistant during many complex field operations. But unknown to us he was in the early stages of a chronic degenerative disease that was beginning to affect his temperament and judgement. We soon lost a valued colleague when he was incapacitated not long afterwards, and died in his early fifties. Hanson Blatz was head of the radiological physics section of the laboratory and preferred to remain in that specialty, rather than assume leadership of the laboratory. Harry LeVine was a superb and imaginative instrument designer and insisted that his instruments be used properly, but had little interest in the data collected. For him the design and operation of a good radiation detection instrument was an end in itself. John Harley had been responsible for the radiochemical part of our programs, and had recently made major contributions to the field studies. Of the four branch chiefs he came the closest to being a suitable successor, but he was not ready to serve as laboratory director. It was decided, in consultation with the branch chiefs, that a director should be selected from outside the organization to serve as a "caretaker" on the assumption that he would eventually be succeeded by Harley. S. Allen Lough, who was a physicist in the New York City Department of Hospitals, was selected to fill the position of deputy and soon became director, when I found two positions too demanding and became the full-time NYOO manager. Shortly thereafter Lough joined the staff of the Division of Biology and Medicine in Washington and was succeeded by Harley as HASL director. Harley served with great distinction for twenty years until he retired.

Renewing Japanese Connections

Upon my return I found the laboratory was in high gear although many of the senior staff were still in the Marshall Islands, where I had left them with the responsibility for salvaging what was left of Operation DUMBO, the unsuccessful attempt to map the close-in fallout fields from multi-megaton weapons, using styrofoam rafts as floating platforms.

Many types of samples, including gummed film, soils, urine, and food were flowing into the laboratory from various parts of the world. John Harley was indispensable as chief of the analytical group, which was heavily involved in developing new techniques of analysis and adapting them to the large numbers of samples sent to us. Harley built a small but extremely talented group that later made the laboratory an international center for training radiochemists and developing standard analytical procedures.

During the remainder of 1954 I was in regular contact with the Japanese concerning the *Fukuru Maru*. In late March, while I was still in Tokyo, Tsuzuki had accepted an invitation from the International Red Cross to attend a conference in Geneva, where he was to deliver a paper on the condition of the fishermen. (For this he was criticized by many of his colleagues, who thought he should have remained in Japan to continue ministering to the patients who were still in critical condition.) When he told me of his plans I suggested he return through the states, by which time I would have returned to New York and could arrange for him to visit several research centers where I was certain he would be a welcome lecturer. He replied that because he was on the purge list of former officers of the Imperial Navy, he doubted that he could obtain a U.S. visa. I said nothing more about the matter at the time and he departed for Europe a few days later. However, I did inform Ambassador Allison who arranged for Tsuzuki an official invitation from the State Department to visit the United States on his way back to Japan. Although I never told Tsuzuki about my role in the matter, he surmised that the invitation was the result of my intervention and made several references to it in later years.

Tsuzuki had been trained in thoracic surgery at the University of Pennsylvania in the 1920s but had not visited the U.S. since then. I met him at the airport in New York when he arrived, and he stayed for a few days at our home in Rockville Centre, Long Island. As we strolled about the neighborhood on the afternoon of his arrival, he expressed amazement at the size of the homes and the well-cared-for lawns. This was an understandable reaction, considering the diminutive size of most one-family homes in Japan. He was also shocked, and I think a bit embarrassed, as when he came downstairs the first morning and found me polishing my shoes! This would have been unthinkable in Japanese society.

I took Tsuzuki to the Brookhaven National Laboratory on Long Island and introduced him to many of the leading medical scientists in the New York area. We then went to Washington where he was given an official dinner by the State Department. On his way back to Japan he stopped to visit Stafford Warren, who was the founding dean of the newly established medical school at the University of California at Los Angeles. Warren had been in charge of health and safety for the Manhattan Project, and had led the first U.S. survey group into Hiroshima after the surrender in September 1945. At that time Tsuzuki, who was the senior medical officer in the Imperial Navy, surrendered his samurai sword to Warren in accordance with proper military protocol. During Tsuzuki's visit in 1954, Warren showed him the sword, which was in a display case, and offered to return it to him. However, Tsuzuki declined to accept it.

Japanese scientists had been cut off from developments in the fields of nuclear energy and radiobiology and were initially uncertain how to undertake fallout studies. General MacArthur forbade them to do research in these fields and had even gone so far as to destroy the only cyclotron in Japan. The radiation measuring equipment in Japan, even as late as 1954, was of prewar vintage. They were not permitted to purchase radioisotopes on the world market, but obtained them by collecting rainwater after U.S. or USSR tests of nuclear weapons. They had no lack of resourcefulness, and in the months following my departure in April had undertaken many ingenious experiments and had even sent an oceanographic research vessel, the *Shinkatsu Maru*, on an expedition to the Marshall Islands to study the movement of fission products from the Bikini lagoon.

Relationships with the Japanese scientists continued to improve during the summer of 1954. It was decided to hold a joint radiobiological conference in Tokyo during the month of November. John Harley and I were members of the U.S. delegation that included a soil chemist, a marine biologist, and a geneticist, as well as, for reasons I never understood, the head of the AEC headquarters Office of Public Information, who headed our delegation. The conference was an immense success in cementing relationships between the environmental scientists of the two countries that have lasted to the present. In one of the highlights of the conference, Harley and I presented our Japanese counterparts with air samples and, most important, the first scintillation counting equipment ever used by the Japanese.

During the weeklong conference, Y. Hiyama reported the findings of the research cruise of the *Shinkotsu Maru*. There was still an enormous reservoir of residual radioactivity in the lagoon of the Bikini atoll. It was seeping into the Equatorial Current at such a rate that it would in all probability be carried westward until it reached the Kurishio Current in the western Pacific Ocean, with which it would be carried northward to Japanese coastal waters. This was an opportunity to learn much about the rates of dilution of contaminants introduced into the western Pacific Ocean, and the rates at which the various nuclides would accumulate in the marine biota. It would also provide information about the levels of radioactivity to which the Japanese and other East Asian people would be exposed from the consumption of seafood. Although the level of exposure was expected to be small, it was important to have data on the subject rather than to leave it to surmise.

Harley and I decided that since there was no plan for a second Japanese cruise, we should send an expedition to the Marshall Islands to track the radioactive plume and study its characteristics. Upon returning home we arranged for the loan of a Coast Guard cutter and its crew, installed a laboratory aboard, and assembled a small team of scientists headed by Harley. The expedition, known as Operation TROLL, left San Francisco on February 25 for a three-month cruise, during which samples of seawater and biota were analyzed for their levels of radioactivity. When the ship reached Japan, Harley provided the Japanese scientists with the information. As expected, the levels of radioactivity in the seafood did not represent a significant health hazard.[1] However, the data collected were useful because they provided new information about rates of mixing in the Pacific Ocean.

Natural Radioactivity in the United States

It is hard to believe that as late as the mid-1950s comparatively little was known about natural radioactivity. This has remained a neglected field of research. Not until about 1984 was the significance of radon as a source of human exposure appreciated. In 1957 we measured external background radiation; we knew that it varied from place to place, partly because the cosmic ray component increased with altitude and partly because of differences in the uranium and thorium content of the rocks and soils. However, the normal range of natural variation had not been investigated.

HASL decided to undertake the first national survey of external radiation exposure from natural sources. Leonard Solon, a physicist on the staff was provided with radiation measuring equipment and a rented station wagon. He and his family travelled crosscountry and back by the northern and southern routes making measurements with a pressurized ionization chamber designed for the survey. He found that the average dose received from external radiation was somewhat less than had been thought, with a range of 70 to 175 mrem per year.[2] In later years the ability to make such measurements was greatly refined, and with the advent of gamma spectrometry in the late 1950s it became possible to apportion the external radiation exposure among its several sources, such as potassium (which has a radioactive isotope, K-40), cosmic rays, radium, and other normally occurring radioactive elements.

Manager of Operations

Apart from HASL, NYOO was an interesting office with which to be associated. There were several AEC operations offices, each with a manager to whom a considerable measure of responsibility was delegated. NYOO was unique in that all other offices were located at the major research or production facilities at Hanford, Oak Ridge, Chicago, Savannah River, Los Alamos, and Idaho Falls. In contrast, the NYOO responsibilities were more regional in nature. We had administrative responsibility for one major facility, the Brookhaven National Laboratory, but the largest part of our responsibility was to administer hundreds of research contracts with almost every university in the Northeast and many industrial companies. Our function was to prepare the contracts, execute them, and see that the research was conducted according to the agreed scope of work. Although the research goals and approval of the individual contracts were usually the responsibility of the Washington headquarters, NYOO was involved with many challenging problems in the administration of these large scientific programs. Two that took a good deal of my time were the Atom Bomb Casualty Commission in Japan, and the construction of the proton accelerator and stellarator at Princeton University.

Atom Bomb Casualty Commission

Within a few months after the atom bombing of Hiroshima and Nagasaki, President Truman authorized a long range study by the National
Academy of Sciences of the delayed effects of the bombings on the survivors. The organization established to conduct the studies was at first
called the Atom Bomb Casualty Commission, and was known worldwide
as the ABCC until its reorganization in 1975 as the Radiation Effects Research Foundation (RERF), jointly supported by the U.S. and Japan.

Financial support of the program was initially provided by the AEC
through the Division of Biology and Medicine, and administration of the
contract was assigned to NYOO, with me as contract administrator.
Shields Warren, director of the Division, asked in 1950 that I visit Japan
with Willard Machle who was on the staff of the Academy and Ernest
Goodpasture, a well-known pathologist at Vanderbilt University who
was also a member of the AEC Advisory Committee on Biology and
Medicine.

A problem that concerned Warren was that the National Academy of
Sciences, which was a smaller organization than it is today, did not have
experience operating its own research programs, and that it might be difficult for them to manage a large research organization of several hundred employees, seven thousand miles away in a country that had not yet
recovered from the effects of the recent war. Machle, Goodpasture, and I
were asked to review the program in Japan and "recommend what
should be done about it."

The three of us arrived in Hiroshima in December 1950 and were billeted in a compound maintained by Australian and British occupation
troops in nearby Kure. We now know that the ABCC has evolved into an
important research organization that to this day, more than forty years
later, has been providing important information about the delayed effects
of radiation on an exposed population. But in 1950, although it was five
years after the bombings and three years after the NAS assumed responsibility for the investigations, the future of the study was in doubt. The nature of radiation effects are such that it would be necessary to continue
the studies through two or more generations at huge cost. It was by far
the largest and most expensive epidemiological study ever planned and it
faced formidable logistic problems. To make things worse, the Korean
war was in progress, and things were going badly for the United Nations

forces. This resulted in serious morale problems among the occupation forces as well as the ABCC staff. No long-range research plan existed, and there were no seasoned research administrators with the project. Morale was not helped by the fact that coal was in short supply and the laboratories so cold that the staff worked in sweaters and overcoats from October to April.

We remained in Japan through the year-end holidays. Hiroshima is now a modern city with beautiful parks and skyscrapers in place of the temporary buildings in which people lived and worked when I first visited the city. Many of the people still lived in shacks and were dressed very poorly. Despite their poverty they nevertheless donned the traditional kimonos on festive occasions, so that I had the opportunity to see the Japanese people before they adopted western forms of dress, an experience I am glad to have had. I was impressed wherever I went by the charm and pleasantness of the Japanese people. Not only did they prove to be cheerful in adversity, but also adaptable in their relationships with others. There was not a trace remaining of the fierce anti-Americanism of the war years.

I also visited the small ABCC staff in Nagasaki, and to this day I have a vivid picture of the medical school, once one of the finest in Japan, but in 1950 still a bombed-out shell, with scrawny goats grazing on the lawns. I spent New Year's Eve riding almost all night between Nagasaki and Hiroshima in a railroad coach in which I was the only Caucasian. At first I felt uncomfortable because the car was crowded with passengers and I was attracting obvious attention. Before long an elderly lady came to my aisle seat, bowed, and offered me a tangerine, which in Japan are surely the sweetest in the world. When I accepted it she seemed very pleased. Other passengers soon followed her example and presented me with bits of food which I accepted with as much grace as I could muster, although I knew not what some of the offerings were.

Machle, Goodpasture, and I spent New Year's Day walking in the hills around Hiroshima where the rice harvest was in progress. We discussed our impressions of ABCC and I was surprised to hear that my two colleagues were of the opinion that the project should be gradually terminated. Both were convinced that the hardships of working in Japan would discourage able scientists from participating in the studies. Moreover, they did not believe the findings would justify the enormous financial burden that the U.S. would be required to assume for many years.

This was the start of a two-year period of uncertainty for ABCC. I was in no position to judge the merits of their arguments on scientific grounds, but an event that occurred in Tokyo on my way home had an important effect on my thinking, as well as on the future of ABCC. I called on General Crawford Sams, who was the senior medical officer on the staff of Gen. Douglas MacArthur. I learned that Goodpasture had met with MacArthur when he passed through Tokyo a few days earlier, and had expressed his somewhat pessimistic views about the value of ABCC. MacArthur had later told Sams that it was important for political reasons to continue the ABCC studies because if they were terminated, it would create a scientific vacuum into which investigators of uncertain scientific credibility would be drawn. Moreover, some of the investigators might be so influenced by political factors as to affect their scientific objectivity. Sams stated emphatically that he hoped ABCC would be continued.

This was the reason why ABCC (now the Radiation Effects Research Foundation) survived those days of doubt and became an important center for the study of the delayed effects of radiation. In early 1951, the advisory committee to the AEC actually recommended that the research be terminated,[3] but it was continued because of the political considerations identified by MacArthur. The information that has been gathered in the two cities now contributes uniquely to our understanding of the delayed effects of radiation and serves as the basis of many of the radiation protection standards. I returned many times to Japan until, by 1956, the ABCC was firmly established in modern laboratories and clinics atop Hijiama hill, overlooking Hiroshima. It has been an immensely successful project, respected worldwide for the quality of its research.

Nearly thirty years later, in 1984, Japanese scientists invited me to return to Japan for a series of lectures. Hiroshima was included in my itinerary; the new laboratory I had last seen in 1956 had begun to show signs of wear but the city itself sparkled with skyscrapers, shopping malls and parks. Modern Hiroshima is an impressive reminder of the resilience of the human race.

Accelerators and Stellarators

During the period when I was manager of the New York Operations Office, there were major advances in the construction of machines that produce nuclear reactions by electrical and electromagnetic means. The design and construction of accelerators and stellarators was a major part of the research program we administered.

The accelerators have been appropriately called "atom smashers" because their basic function is to study the structure of atoms by splitting them into their constituent parts. The first cyclotrons and linear accelerators, built in England and the U.S. before World War II, contributed in a major way to our understanding of atomic structure. They were also the first sources of small quantities of artificially produced radioactive isotopes. But the first machines were of modest dimensions and were constructed in university laboratories by the scientists who designed and used them.

After the war, the high-energy experimental physicists who worked with these machines were remarkably successful in attracting huge sums of money for the construction of machines at the Brookhaven National Laboratory and on university campuses. Brookhaven built the cosmotron, the first of the large postwar proton accelerators, which produced a beam of three billion electron-volts. The Brookhaven Laboratory was intended to serve as a center where such expensive highly complex machines could be built, constructed, and operated, as a service to all universities in the Northeast.

However, some of the larger universities wanted their own machines. When I became manager of operations I found that Princeton was in the midst of two huge projects. These involved construction of a proton accelerator, and a classified program that was attempting to create energy by the fusion of hydrogen, known by the code name SHERWOOD. Astrophysicists understood the processes by which fusion reactions take place in the stars and were attempting to produce controlled fusion in the laboratory as the first step towards developing a system of energy production based on the virtually unlimited supply of hydrogen.

The principle, on which SHERWOOD and most subsequent programs of controlled thermonuclear reactions have been based, is that if a stream of light-element ions is confined by a magnetic field and heated sufficiently in a radio-frequency field, the atomic nuclei combine (fuse), a process that releases significant amounts of energy. However, the fusion reactions also release copious quantities of neutrons, which can be used to irradiate uranium and produce plutonium, the artificial element used in nuclear bombs. Because of the possible military applications of the SHERWOOD program, it remained highly classified until 1958 when it was decided that it was not a practical method of plutonium production.

The Princeton system required confinement by magnetic means of the hot ions (plasma) within a toroidal structure called a stellarator because of the astrophysical origin of the concept. The engineering problems in this design remain formidable to this day because of the difficulties in confining the plasma long enough to permit the required thermonuclear reactions to take place. Nevertheless, the importance of controlled thermonuclear reactions as a potential source of cheap, clean energy is such that research programs have continued in many other laboratories in the U.S., as well as the Soviet Union and other countries.

When I was first introduced to SHERWOOD in 1954 I was so impressed with the engineering problems that I insisted that Princeton invite industrial support. Scientists were accustomed to building the experimental apparatus used in their research but it was time to seek the assistance of industry in solving the difficult construction problems of the stellarator— to design and fabricate huge magnets, maintain high vacuum conditions within a large torus, and eventually to design a system for removal of the excess energy it was hoped would be produced. I found a good deal of resistance to bringing industry into the SHERWOOD program. The capable scientists involved in the program saw no need to seek the help of industry to do things they had traditionally done by themselves. In addition to the stellarator, AEC was also financing construction of the large proton accelerator at Princeton. I pointed out to the university officials that the government was spending more money on construction of the new Princeton facilities than the university itself had invested in the entire campus over the previous two hundred years. I finally convinced the university officials and project staff of the necessity of obtaining industrial support; the RCA Laboratories, located not far from Princeton, became a partner in the stellarator program.

Despite my belief in the competence of the Princeton scientists, I could not share their optimism that they were on the verge of building a prototype stellarator capable of producing electrical energy. In the mid-1950s the project submitted a proposal seeking AEC support for the design of a 100-megawatt thermonuclear power station, large enough to supply the electrical needs of Elizabeth, New Jersey. The AEC did not accept the proposal, which would be naive even today, more than thirty years later.

During the summer of 1959, just prior to my retirement from the AEC, I spent several weeks at the University of Mexico lecturing on the subject of environmental radioactivity. On my last evening in Mexico City, a farewell dinner was held for the U.S. scientists who had spent the summer with our Mexican colleagues. After dinner there was some general discussion of a number of subjects, including the future of controlled thermonuclear energy. I was asked when the first electricity would be produced and I replied "from 25 years to infinity." At this writing, thirty years later, I believe that is still a valid estimate. Great progress has been made, but the basic problems remain. The ultimate feasibility of the method has not yet been demonstrated.

The Nuclear Ship Savannah

One of the most interesting responsibilities of the NYOO was to design and build the first nuclear-powered merchant ship, named the *Savannah*, after the first steam-powered vessel to cross the Atlantic. Its construction was proposed by President Eisenhower in April 1955 as part of his Atoms for Peace program. It was his plan that the *Savannah* would travel the world, bringing messages of peace from the American people.

The first of the nuclear submarines, the U.S.S. *Nautilus* had already slid down the ways in Groton, Connecticut in January 1954. There was optimism that the U.S. merchant fleet could also be powered by nuclear energy because the pressurized-water reactors used in the submarines had demonstrated many advantages over oil. A contract to design the power plant was awarded to the Babcock and Wilcox Company, and the vessel was constructed in the Philadelphia yards of the Sun Shipbuilding Company. The *Savannah* was launched in the summer of 1959 by Mamie Eisenhower amid great optimism. Because I was contracting officer for the government, I, and Irma, participated in the festive occasion.

Unfortunately, the *Savannah* was one of the first targets of opposition to the civilian use of nuclear energy. Some cities abroad refused to allow the vessel to enter their harbors. In New York, the *Savannah* could not sail majestically past the Statue of Liberty to its Manhattan pier because local officials would not allow the reactor to operate, forcing it to be towed ignominiously through the harbor and up the river. To add to the woes of the *Savannah,* serious labor problems developed because the maritime union was unwilling to accept many of the changes in job classifications and assignments necessitated by the ship's power plant. These problems, together with the lack of public acceptance, ultimately resulted in the ship being decommissioned in its home port of Savannah, Georgia

—but not before Joe Clarke, Hanson Blatz, and our colleague from Columbia, Harald Rossi, enjoyed an opportunity for an all-day ocean cruise on it.

International Activities

An important part of my duties as head of NYOO was to represent the AEC in international activities. I assumed the position of manager at a time when the AEC was being involved increasingly in foreign affairs. Much of this involved NYOO because (1) New York was the headquarters for the United Nations, (2) there was increasing worldwide concern about the fallout from weapons testing then in progress, (3) HASL was located in the NYOO and was operating a worldwide fallout measuring network, and (4) there was international interest in President Eisenhower's Atoms for Peace Program. I had been introduced to foreign activities in my experiences with the Atom Bomb Casualty Commission and the Lucky Dragon affair, both of which involved trips to Japan and correspondence with Japanese scientists.

In 1955 there were two major international developments in which HASL played important roles. These were the First United Nations Conference on the Peaceful Uses of Atomic Energy, and the formation of the United Nations Scientific Committee on the Effects of Atomic Radiation, ever since known as UNSCEAR.

By early 1955, plans were underway for the United Nations Conference on the Peaceful Uses of Atomic Energy, to be held in September at the Palais des Nations, the UN headquarters in Geneva. Several such conferences have since been sponsored by the UN, but none have engendered the excitement of the one held in 1955.

The conference resulted from a sequence of events that began with a speech by President Eisenhower at the United Nations in December 1953. He proposed that nuclear weapons in the stockpiles of the great powers be withdrawn gradually, stripped of their military casings and... "adapted to the arts of peace." I was present on that exciting occasion and was impressed by the enthusiasm with which the proposal was received by all delegates, including those from the Soviet Union, who jumped to their feet to join in the standing applause. That talk, which came to be known as Eisenhower's "Atoms for Peace Speech," seemed for a while to be a turning point in relationships with the Soviets and resulted in plans for the 1955 Geneva conference. It was a huge affair,

Industrial hygiene training class and faculty, Liberty Mutual Insurance Company, Boston, 1940. *Seated, left to right:* Frank Shear, Stuart Gurney, Charles R. Williams. *Standing:* Harlow Chapman, author, William Davis, Stan Ballard, Joseph Houghton

With Hiroshima children in 1950. The city in the background is being rebuilt, five years after the bombing.

Conference with the committee of Japanese scientists appointed to investigate the fallout on the *Lucky Dragon*, March 1954. To the author's left are William Leonhart, First Secretary, U.S. Embassy and John Morton, Director of the Atom Bomb Casualty Commission in Hiroshima. Across the table, are *(third from left);* Masanori Nakaidzumi and *(third from right)* Kenjiro Kimura.

At the First International Conference on the Peaceful Applications of Atomic Energy, Geneva, 1955. *Left to right:* the author; Detlev Bronk, President of the National Academy of Sciences; Gioacchino Failla of Columbia University; Walt Whitman of MIT; and Professor Fujioka of Japan.

The U.S. delegates to the first meeting of the United National Scientific Committee on the Effects of Atomic Radiation (UNSCEAR). *To the left*, Shields Warren, chief delegate, with Austin Brues and the author as alternates.

The first UNSCEAR meeting in December 1955. U.N. Secretary General Dag Hammarskjold occupies the left-hand position at the far table. The author is behind and to the left of Shields Warren, the chief U.S. delegate.

With Masao Tsuzuki, Masanori Nakaidzumi, and Rolf Sievert, 1958

At a 1965 Brookhaven
National Laboratory
conference with Lewis L.
Strauss and Willard F. Libby,
then AEC commissioners.
Both played important but
controversial roles in the
fallout debates that began in
the mid-1950s.

With Edward R. Murrow in a 1958 interview on fallout for his popular
television program.

Guests of Queen Fredrika of Greece, during her 1958 visit to New York. *Left to right:* Mrs. John McCone, wife of the AEC chairman, Leland Heyworth, Irma, Chairman McCone, Fredrika, and the author with his back to the camera. (Joe Scherschel, *Life Magazine*, Time-Warner Inc.)

With Eduardo Penna-Franca and Father Thomas L. Cullen at a 1965 Pan-American Health Organization conference in Washington.

At a 1965 industrial health conference with three long-time associates, *left to right:* John H. Harley, Norton Nelson, and David Goldstein.

The U.S. Public Health Service Environmental Radioactivity Advisory Committee. *Left to right, front row:* John H. Rust, the author, James G. Terrill, and Dr. Thompson. *Rear row:* John W. Healy, Richard F. Foster, Edward B. Lewis, Julian H. Webb, Charles Weaver, and John W. Gofman.

Father Thomas L. Cullen assists in the formalities at the 1971 award of an honorary doctorate by the Catholic University of Rio de Janeiro.

AEC Chairwoman Dixy Lee Ray presents the author with a citation and gold medal in 1974, fifteen years after his retirement from the agency.

CHEMICAL AND ENGINEERING
NEWS
June 24, 1957

Experts Get
Together For
Congress, Can't
Agree on Hazards
from Fallout . . . 16

Panel assembled by the
Congressional Joint Committee
on Atomic Energy in 1957 to
discuss the health implications
of radioactive fallout. The
composition of the panel
illustrates the variety of
specialists involved in the
studies. *Left from the
stenotypist*, Lester Machta
(meteorology), William F.
Neumann (bone metabolism),
Austin M. Brues (radio-
biology), Charles L. Dunham
(biomedical research
administration), James V. Neel
(genetics), Shields Warren
(pathology), Willard F. Libby
(chemistry), the author, Bentley
Glass (genetics), and Wright
Langham (physiology).
American Chemical Society

In a Colorado mine during a
1967 experiment with an
electrostatic precipitator
designed to remove radon
daughter products from the
atmosphere.

New York City Mayor John V. Lindsay addresses a 1969 assembly of the Sanitation Department officers. Commissioner Griswold Moeller relaxes with his arms crossed while the author reviews his notes next to the union president, John DeLury.

Judge Edward C. Maguire administers the oath of office to the new president of the Board of Water Supply of the City of New York, April 1969.

On the monazite sands of
Kerala, in southwest India
during a 1980 visit. The people
who live in this area are
exposed to abnormally high
levels of natural radioactivity.

With Irma and Lin Bin, director of the Chinese National Cancer Institute,
during a 1981 visit to a hospital in southern Yunnan Province. We were the
first Americans to visit the area in thirty-six years and were invited to
investigate a serious epidemic of lung cancer among tin miners.

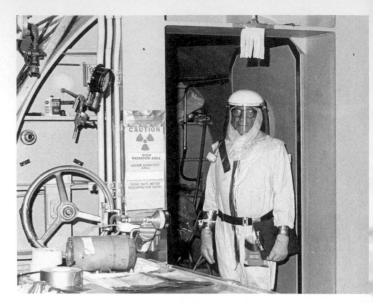

As a member of the Three Mile Island Safety Advisory Board, entering the reactor containment building during an early stage of the cleanup.

Irma and the author join Warren K. Sinclair and Lauriston S. Taylor, the president and former presidents of the National Council on Radiation Protection and Measurements, and wives, at a 1983 reception following the author's delivery of the Seventh Annual Taylor Lecture at the annual NCRP meeting.

The 1987 reunion of World War II-era industrial health specialists. The conference was sponsored by the Chemical Industry Institute of Toxicology in Chapel Hill to identify groups of workers who were heavily exposed to toxic chemicals, and for whom epidemiological studies might be important to determine if delayed health effects developed. *Seated, left to right:* James H. Sterner, the author, Warren A. Cook, Leonard J. Goldwater, and Jacqueline Messite. *Standing:* Fred J. Viles, Arthur C. Stern, Irving R. Tabershaw, Hervey B. Elkins, W. Clark Cooper, Lewis J. Cralley, William R. Bradley, Ronald Buchan, Henry N. Doyle, J. Wister Meigs, and Harry F. Schulte.

The author with *left,* Nicholas Lizzo and *right,* Scott Morris, the first two recipients of the Eisenbud graduate fellowship established by former students and friends. The portrait is of Anthony J. Lanza, who founded the NYU Institute of Environmental Medicine in 1947.

Our three boys, with their wives and our grandchildren, assemble in Bermuda in 1989 to celebrate our fiftieth wedding anniversary. *Left to right:* (all Eisenbuds), Debbie, Frederick, David behind Lynne who is holding Benjamin Joshua, who is in front of Irma, Elliott, Jennifer, behind Lauren, the author, Barbara, Michael, Chad, and Daniel.

attended by 3600 scientists from all the countries of the world. In preparation for that conference, the nuclear powers declassified vast quantities of information. In addition to the hundreds of papers presented, there were elaborate scientific exhibits.

HASL played a conspicuous role at that conference. Because the laboratory had been at the forefront of the development of equipment for the measurement of radioactivity, we were asked to assemble an exhibit. Harry LeVine played a major role is designing and assembling the exhibit, which was a huge success. Not only was it attractive and informative, but its location in a robin's-egg blue trailer on the lawn in front of the palais and its ample stock of Coca-Cola made it a popular exhibit, especially to the Soviet delegates.

The trailer did not get to Geneva without difficulty. It was originally planned that it would be in the exhibit hall but it was too big. When it was suggested that it be on the front lawn, the State Department was unsupportive. However, AEC Deputy Director of Biology and Medicine John Bugher, who was always a strong supporter of HASL, contacted the World Health Organization through Dr. Coigny, the WHO representative at the UN in New York. Thanks to his support the trailer was accommodated in the exhibit, to the credit of all concerned.

The highlight of the conference, so far as I was concerned, was my presentation before a plenary session of the paper I prepared jointly with Joseph Quigley who had been my medical assistant at HASL, on the "Industrial Hygiene of Uranium Processing." The paper summarized seven years of work by our laboratory and was the first publication on the subject.[4]

Prior to the 1955 Geneva Conference, there was little familiarity with nuclear technology outside of the countries that pursued its military applications. The conference provided the world with the basic information needed to develop nuclear power and also stimulated interest in the use of nuclear techniques in research and medicine. During the months following the conference, New York became a major center for visits of foreign officials who wanted to learn more about nuclear energy. New York was an attractive place to visit because the Brookhaven National Laboratory, more so than others, was engaged in unclassified research in both the biological and physical sciences. In addition, HASL was well equipped to show foreign visitors the basic instrumentation used in research laboratories and could also demonstrate the techniques used to measure fallout from the nuclear weapons testing programs that were being conducted by the U.S., U.K., and USSR. Over a period of about two years, we were visited by Queen Fredrika of Greece and her children, the President of Italy, and Haitian dictator "Papa Doc" Duvalier, to name a few.

Fredrika was a particularly attractive and intelligent woman, a grand-daughter of Queen Victoria and a mystic who was trying to find the answers to the mysteries of life through atomic energy. She invited Irma and me to a small luncheon in her suite in the Waldorf-Astoria Tower, which was also attended by the chairman of the Atomic Energy Commission, John McCone, and the Brookhaven Director, Leland Heyworth. A day or two later, Irma and I drove with her to Brookhaven where she enjoyed a tour of the laboratory and was feted at a dinner in her honor.

Fredrika practiced none of the formalities expected of a queen. Shortly before she left New York, I received a call from her asking if I would return to her suite later that day for an informal recital by Gina Bachauer, a well-known Greek pianist, with whom Fredrika had spent the war years in North Africa as a refugee from fascism. When I arrived at her Waldorf suite, the queen and the pianist were debating the proper location for the piano which Fredrika, in her stocking feet, pushed into the position they chose. Her mysticism led her in subsequent years to a religious refuge in India, where she died at an early age.

UNSCEAR

The United Nations Scientific Committee on the Effects of Atomic Radiation (UNSCEAR) had its origin in the Joint Congressional Committee on Atomic Energy, the executive director of which was Corbin Allardyce, who had formerly been on the staff of NYOO. One day in early 1955, he and an assistant dropped in to see me and asked about the value of a United Nations committee to collect and evaluate information about environmental radioactivity. I responded very positively, as a result of which I was invited to Washington to discuss the matter with Representative Sterling W. Cole, Chairman of the Joint Congressional Committee on Atomic Energy. Cole was a close friend of Lewis Strauss, who was not only Chairman of the AEC, but also the president's advisor on atomic energy matters. The idea quickly caught on, and was presented to the General Assembly of the United Nations by Ambassador Henry Cabot Lodge. As a result, the United Nations Scientific Committee on the Effects of Atomic Radiation was established and held its first meeting in December 1955 in New York. It was initially a fifteen-country committee, with Shields Warren, professor of pathology at Harvard, as U.S. delegate, and Austin Brues, Director of the Biology Division at Argonne National Laboratory and me as alternate delegates.

At the first meeting Warren made the important suggestion that the committee confine itself to assembling and summarizing the available scientific information about the biological effects of nuclear radiation. UNSCEAR should make no recommendations. This policy was adopted, making it possible to avoid the inevitable bickering over politically sensitive subjects, such as weapons testing. UNSCEAR has been successful in limiting itself to reviewing the world literature and assembling the information issued every few years in reports which are classics in international scientific collaboration. The recommendations have been confined to other groups, such as the ICRP, its various national counterparts, and the national regulatory bodies.

I remained an alternate U.S. delegate to UNSCEAR until 1960 when I began to find the work of the committee burdensome. Its headquarters were in New York, only a few blocks from my laboratory at New York University, which made me too accessible. Since much of the work of UNSCEAR was concerned with environmental radioactivity, I found myself overly involved with the work of the secretariat. John Harley took my place on the delegation and has remained in that position for many years.

Although the work of UNSCEAR proceeded smoothly, with relatively few problems due to East-West friction, there were a number of incidents worth noting, some of which serve to show how silly international relationships can become. One of the responsibilities assumed by the committee for its first report was the preparation of a map of the world on which would be recorded the fallout measurements made by the various countries, each of which was requested to submit its measurements to the committee. In this the Soviets cooperated fully and submitted not only data collected within their own boundaries, but also data from the People's Republic of China, with which the USSR had been collaborating. However, China had not yet been admitted to the United Nations, primarily because the U.S. continued to recognize the Taiwan government and to block the entrance of mainland China. The State Department therefore prevented use of the Chinese data by the committee, so the report was published with a map showing China as a vast blank. Fortunately, this was one of the very few ways in which political considerations influenced the scientific output of UNSCEAR.

There were also opportunities for levity. From the first meeting, English increasingly became the unofficial working language, although there were simultaneous translations from and into Spanish, Russian, and French. Many of the delegates who spoke no English when we first met in 1955 were working hard to learn the language, and gradually became so proficient that we frequently conversed without having translators present. However, the French delegates were instructed to speak in their own language during the formal sessions. On one occasion when the delegates, who were seated alphabetically around a long table, were voting, the Australian, Belgian, Brazilian, and other delegations each gave its approval with a quick "yes." The yes-voting process continued through the French delegate who also said "yes," only to exclaim a few seconds later, "Main non, le délégué de la France dit oui!"

In 1956 Israel and Egypt became engaged in the Suez War. It started when the committee was meeting. Irma and I had invited the delegates to our home for a Saturday reception and dinner. A Security Council meeting about the outbreak of hostilities was scheduled for that evening, and the Soviet delegate Professor Lebedinsky told me he regretted that his group would be unable to attend our party because the Russian translators would be on duty at the Security Council meeting. We were still totally dependent on translators when communicating with our Soviet colleagues. (The situation has since improved markedly, far more because the Soviet scientists have learned English than vice-versa.)

The reception had just begun at our home on Long Island when Lebedinsky and the full Soviet delegation emerged unexpectedly from two limousines, but without their interpreters. Except for a smatter of French, the Soviet delegates spoke only Russian. Irma, resourceful as always, remembered that she had a friend in town who taught Russian, and we were fortunately able to locate her at another party which she obligingly left to come to our rescue. It was a successful evening, in which I am sure that our Soviet colleagues were pleased to be free of their official interpreters, who we always suspected fulfilled other duties. I have been told by knowledgeable people at the U.N. that it was the first time that any unaccompanied Soviet delegates had attended a social affair in the United States.

The years 1956 through 1958 were ones in which there were repeated

opportunities for interesting overseas assignments. After receiving foreign visitors to HASL, and also being involved in their trips to the Brookhaven National Laboratory, I received invitations to visit European and
Asian countries that were developing programs for the peaceful applications of nuclear energy. I also visited many of those countries as part of
the "Atoms for Peace" program.

U.S. policy included gifts of small reactors that were valuable for
biological and physical research, radiation detection instruments, and
instruction in special courses. There was also increasing international
interest in fallout from weapons testing, so that a few of my trips were to
discuss methods of measuring fallout, as well as the public health significance of the data being collected worldwide by HASL.

During this period I found myself being drawn into a close relationship
with Lewis L. Strauss, Chairman of the AEC and the White House advisor on atomic energy matters. Strauss, among the first of the extreme
conservatives in the Republican party, was a controversial figure during
most of his tenure. He was also a military hawk as suspicious of Soviet
intentions towards the United States as any official in government during
that coldest period of the cold war. He was not popular with the majority
of U.S. scientists for many reasons, among which was his role in the dis-
loyalty hearings against J. Robert Oppenheimer in 1954. Strauss was an
independently wealthy man who was in his mid-fifties and at the height
of his career. There was very little reason why he should have befriended
me, since my political views were generally liberal, quite the opposite of
his. Moreover, he was aware that I thought he and other members of the
Administration had badly handled the matter of the CASTLE fallout.
Nevertheless he treated me with great kindness, and may have been responsible for the solid position I occupied in the AEC, despite the fact
that I had alienated many senior officials because of my persistent efforts
to study the effects of fallout from nuclear weapons testing and make the
information available to the public.

One day in May 1957 I received a call from Strauss, who told me that
"he and the president" wanted me to undertake a special assignment that
might require me to be in Europe for a few days. I didn't believe that the
president was really involved, knowing that Strauss frequently name-
dropped when he wanted something done. I said I was always ready to
assist, and agreed to come to Washington for a conference. When I arrived, I was told that Conrad Adenauer, the widely respected West German Chancellor, was scheduled to come to Washington in late May at

which time he would like to discuss the subject of fallout with me. I returned for that conference, which I attended with Strauss and Secretary of State John Foster Dulles.

Adenauer and I spoke for only a minute or two. He was affectionately called *der Alte* because of his advanced age. He looked every bit the octogenarian that he was, but his eyes were sharp and his mind clear. He asked if I could come to Bonn for discussions with him and his staff, and I of course replied that I would be happy to do so to review both the subject of fallout from weapons testing, and what I knew about the effects of fallout in the event of a nuclear war, which could cover Western Europe with lethal radioactivity.

I left for Bonn in mid-June and was met at the airport on a Sunday morning by a member of the embassy staff who took me to a comfortable hotel with a room that overlooked the Rhine. After a restful day I was taken to the embassy residence where I had dinner with Ambassador David Bruce and a small group of his staff. Bruce, who later served as Ambassador to the Soviet Union, was one of the many truly great men I have known in public life who could have made substantial contributions to our country if they could have been induced to enter politics and compete in the elective process. There have been exceptions, but the best of our citizens do not run for elective office.

Adenauer was in his eightieth year and was subject to respiratory tract infections that kept him in bed. Bruce informed me that the chancellor was in the midst of such a bout and would be unable to see me for a few days. Both the German Foreign Office and the embassy were embarrassed by the fact that I had travelled so far at the chancellor's invitation, only to be unable to meet with him. During the next several days I had several meetings with Foreign Minister von Brentano and many of his staff, as well as scientists from the universities and other branches of the German government for general discussion of the problems of nuclear weapons fallout. The prospect that a nuclear war could blanket Western Europe with lethal amounts of radioactivity was a matter that greatly concerned them, but they believed that the nuclear umbrella was the only way in which a Soviet drive into West Germany could be prevented. Thirty years later that dilemma still existed, but is now hopefully being resolved through the process of Soviet democratization.

The general embarrassment about *der Alte's* indisposition had its benefits so far as I was concerned. My hosts took me on long drives through the countryside, to the opera in Cologne, and on a beautiful trip on the Rhine. After a week of waiting it was necessary for me to return to New York without having had my conference with Conrad Adenauer, whose slow recovery from his latest bout of pneumonia was by then the subject of national concern.

Relations with Congress and the Press

Although the AEC headquarters was in Washington, there were many times when I was called on to represent the agency in dealing with the media, the United Nations, and the Congress.

In 1946 the Congress established the Joint Congressional Committee on Atomic Energy (JCAE), the membership of which contained many leaders from both houses. JCAE was an oversight committee which held frequent hearings concerning whatever subject happened to interest the members at any given time. For thirty years, until the AEC was disbanded in 1977, the JCAE exercised a powerful influence over AEC affairs. During most of its existence, the committee provided valuable support for AEC but there were also many periods when the interventions of the committee proved distracting and disruptive. There were times when the AEC could do no wrong in the eyes of the JCAE and other times when it could do nothing right.

One of the earliest of the confrontational hearings took place during the tenure of the first AEC chairman, David E. Lilienthal. I was not involved in this particular hearing, but I will discuss it briefly. Senator Bourke Hickenlooper, chairman of JCAE, had charged AEC with "incredible mismanagement," initially because it could not account for a trivial amount of uranium 235 that was missing from the records of one of the major research centers. The inconsequential discrepancy was no doubt the result of statistical uncertainties in the assay procedures, but one thing led to another and eventually it was pointed out to Lilienthal that the high rate of personnel turnover within the agency was another example of how the agency was being mismanaged. Lilienthal promised to respond to the accusation on the following day, at which time he reported that an overnight investigation had discovered that most of the turnover was among the young women on the staff, and that the reason most often given for resignation was expectant motherhood. Lilienthal added that while this might be the result of "incredible mismanagement," it was clearly something for which he could not take responsibility. I relate this story to illustrate the extent to which the hearings sometimes involved minutiae, and also to illustrate the importance of a sense of humor in dealing with the Congress at hearings attended by the press. This was a quality I often regretted not having. I always thought of witty things I should have said, but always too late to do any good.

I appeared several times before JCAE to report on our fallout studies and frequently became caught in the political cross fire that often took place. This was a particular problem when Senator Clinton Anderson of New Mexico was JCAE chairman and Lewis Strauss was chairman of the Atomic Energy Commission. Anderson's dislike for Strauss was intense, and it resulted in difficulties for the AEC staff who appeared before him at the frequent hearings. In 1959, Anderson gathered sufficient votes in the Senate to block Eisenhower's nomination of Strauss as secretary of commerce.

My position in the New York Operations Office brought me in frequent contact with the media, much of which was headquartered in Manhattan. In the 1950s, atomic energy was still novel and attracted the attention of the best science reporters and feature writers. Grace Wells Urrows, who was the NYOO information officer, maintained excellent relationships with the reporters so that our office became a valuable resource for them. Grace would arrange interviews, provide them with reports, and make it possible for them to visit AEC facilities. Most of the press representatives we dealt with were conscientious and devoted whatever time was required to study the background material we gave them. However, this changed by the late 1950s when reporting about the health effects of atomic energy began to attract the attention of nonspecialists who were either unable or unwilling to devote the time required for research. As a result, the reporting soon became more sensational than factual, a problem that has persisted.

International Civil Servant

Beginning in 1957, I gave considerable thought to whether I should move more actively into the international arena. Although in retrospect the U.N. has not been successful in dealing with the major political and military issues of our time, UNSCEAR did demonstrate that there were certain areas in which the UN could contribute to international understanding. Moreover, it was then widely believed that nuclear power and the use of nuclear energy in research and medicine would solve many of the world's problems. The 1955 Geneva Conference on the Peaceful Uses of Atomic Energy had been a huge success, and in its wake the General Assembly took the first steps towards establishment of the International Atomic Energy Agency.

My position in the AEC and my familiarity with the scientists of many nations made me a logical candidate for a position with the new agency. It was suggested at an early stage that I should join the staff of the Preparatory Commission as senior scientific advisor. It was something I wanted to do, and over a period of a few months I made quiet efforts to encourage the suggestion, while at the same time taking the position that I would only undertake the assignment if I was asked to do so by the AEC. When the request came, I took a three-month leave to join the staff of the United Nations Secretariat.

Unfortunately, it was a disappointing experience. The Preparatory Commission was composed of ambassadors to the UN from a dozen or so countries who met daily over many weeks to act on drafts of the various reports and position papers that were required for the IAEA to begin its work. Most of the debate concerned procedural matters such as how often the Board of Governors should meet, the amenities for them, what committees of the board should be established, and the relationship of the new agency to the other specialized agencies of the UN, such as the World Health Organization, and the Food and Agricultural Organization. As the senior scientist on the Preparatory Commission I was naturally interested in the scope of the training programs, the laboratories needed, and a much-needed reactor inspection program. Above all, there was a need for international inspections of nuclear facilities to prevent the proliferation of nuclear weapons. But during those dreary Preparatory Commission days, there was little talk about such substantive matters, because the ambassadors were more interested in procedures and protocol.

When the time came to select a headquarters for the agency, it was decided that it should be located in Austria, which was located between the eastern and western blocs of nations. I was included in a small group that visited Vienna to find temporary facilities and develop the terms for an agreement with the Austrian government. Other members of the delegation were the Czech and Brazilian ambassadors to the United Nations,

Brian Urquhart (who was an assistant to Secretary General Hammarskjold), and Paul Jolles, who was directing the work of the Preparatory Commission secretariat. The Brazilian was Carlos Bernardes, Chairman of the Preparatory Commission, who began the trip inauspiciously by missing the plane to Vienna!

Vienna was emerging from the destruction and desolation of World War II, despite which we enjoyed the perquisites of a state visit, as official representatives of the United Nations on a mission to establish the agency headquarters. In between evenings at the opera and official dinners, I toured the laboratories of Vienna, primarily so that I could see what facilities might serve the needs of the IAEA during its startup period, but also because some of the laboratories were world famous, and I was anxious to visit them. In particular, I wanted to visit the once famous Radium Institute, the oldest of several such laboratories in Europe. Among the Austrian physicists who were once on the staff of the Institute were Lise Meitner, co-discoverer of nuclear fission, and two Nobel laureates, Victor Hess, the co-discoverer of cosmic rays, and George de Hevesey, the first to use radioisotopes as tracers in scientific experiments. I couldn't help but be depressed by my visit. There were no young scientists. The facilities were obsolete, and the laboratories were illuminated only by a few dim low-wattage electric bulbs hung by wires from the ceiling.

On my return to New York I drew up the first budget for the IAEA scientific program, but it received a low priority from the diplomats on the Preparatory Commission, who seemed more interested in spending the funds at their disposal for meetings of the Governing Council, international travel, and the amenities associated with a United Nations agency. I realize now that I should not have been so pessimistic about IAEA's future; I simply didn't understand the realities of international politics. The IAEA has just celebrated its thirtieth birthday and is a mature and successful member of the family of specialized agencies of the U.N. It serves an important role in helping to control the proliferation of nuclear weapons and in setting guidelines for the safety of nuclear power reactors. Although I spent only three months with the Preparatory Commission, it was an interesting experience that proved to be far more productive than I realized at the time.

During those three months the question naturally arose as to whether I should remain on the IAEA staff after its formation. I wanted to do so very much, in a senior scientific capacity, and was encouraged by several key people in both the AEC and the State Department. It was apparent that the top positions would have to be distributed evenly between the nations of the eastern and western blocs, but I assumed that someone from the neutral nations would be appointed director general. This turned out not to be the case and, much to my surprise, Sterling Cole, the chairman of the Joint Congressional Committee on Atomic Energy, was named to the position. There were to be two or three deputies appointed, one of whom would be concerned with health and safety matters. I had expressed an interest in that position, but this became out of the question when it was announced that Cole would be the Director General. This was a disappointment to both Irma and me because we had begun to look forward to living in Vienna for a few years.

In 1957 I became involved in yet another series of abortive discussions about international matters, this time on the subject of disarmament. There was an important disarmament conference underway in London at the time of my visit to Bonn, where I had benefitted from several informative conversations on the subject with Ambassador David Bruce. He had been receiving daily cables from the conference and was concerned that the technical advice being received by the U.S. delegation was ad hoc and lacked balance. He was referring to the fact that Nobel laureate E.O. Lawrence had been summoned there a few days before as a technical advisor; Lawrence had the reputation of being hawkish on the subject of relations with the USSR.

A few weeks later while discussing the subject with Gordon Dean, former AEC chairman, I asked whether the U.S. government had a staff that could competently analyze the technical aspects of a sincere disarmament proposal. He did not think so, which was an interesting commentary on the state of the cold war at that time. In August 1957 I had an opportunity to discuss the matter with Lewis Strauss and was surprised at his strongly positive reaction to my suggestion that a small but highly skilled group of scientists be assembled to explore the technical problems that would be faced, particularly with respect to the matter of verification, in the event of a disarmament agreement. In response to his query, I said I would be willing to lead such a study.

Less than a month later, I was talking to I. I. Rabi, Columbia University's Novel laureate physicist, who was chairman of President Eisenhower's Scientific Advisory Committee. When we touched on the subject

of disarmament, I was amazed to hear him say that only a few days before the president had discussed with him the need to establish a group to
study the technical problems associated with disarmament. Since the
words Rabi used were very much like mine, I was certain that Strauss had
relayed my views to the president and I became optimistic that something
might develop out of the suggestion I had made. Within another few days
I was informed by AEC General Manager Kenneth Fields that Strauss
wanted the "disarmament inspection staff" to be assembled immediately,
though it might not be possible to make the arrangements in less than one
month. In fact, nothing happened. A few months later I mentioned the
matter to Strauss again, but he didn't want to discuss it.

Full-time Administrator

By 1957 I needed to relinquish one of the posts I was holding. For three
years, thanks mainly to Joe Clarke's assuming the day-to-day administrative burdens of the manager's office, I had served as both HASL director
and head of NYOO. The responsibilities of both positions had expanded
during those three years, so it was now necessary to relinquish one of the
two.

The position of manager was higher in the AEC hierarchy, and it
seemed more challenging. This was, after all, the heady period of atomic
energy development, which had begun with the Geneva Conference of
1955. As the AEC representative in the northeastern U.S., I was very
much involved in assisting industry to enter the nuclear energy field and
was in regular contact with the officers of many large industrial companies. I thought, erroneously as it turned out, that at forty-two I had progressed as far as I could in my scientific specialty and that I should devote
full time to being manager. In August 1957 the transition was made and
S. Allen Lough became director of HASL.

I was at this time very much concerned with the future of HASL. It was
a unique laboratory in many ways and had earned worldwide respect for
its work in industrial hygiene, health physics, and environmental science.
However, it was still not solidly accepted as part of the AEC because the
agency's policy was to contract with industry or universities to fulfill its
research and production functions. Of the dozens of plants and laboratories involved with AEC research and production, only two were staffed
by AEC personnel. The other was the Raw Materials Laboratory in New

Brunswick, New Jersey which was involved with uranium research. The headquarters staff frequently suggested that HASL be shut down and its work taken over by one of the national laboratories. John Bugher and I kept HASL alive and thriving by constant objections to such a move. We were both concerned about the viability of HASL as an independent laboratory staffed entirely by AEC personnel. Since Bugher was planning to retire from the AEC and return to the Rockefeller Foundation just as I began to think about retiring from the laboratory directorship, he and I discussed other arrangements to assure the future of the laboratory.

Of the various possibilities, the most attractive was to have the laboratory operated under contract with the Institute of Industrial Medicine (now the Institute of Environmental Medicine) at the New York University Medical Center. HASL and the institute were both founded shortly after the end of World War II. In the previous ten years, a close relationship had developed between the two organizations. I held the position of adjunct professor at the institute, as did William Harris and Hanson Blatz; we taught courses there in industrial hygiene, health physics, and air pollution. We had loaned specialized equipment to the institute and had arranged for AEC financial support for its research on the carcinogenic effects of radiation. Norton Nelson, the remarkably able director of the institute had also served as a consultant to HASL, as did others on his staff. In short, there had been a close and productive relationship between the two groups of scientists.

In January 1957, Bugher and I discussed a contract that would make it possible to merge HASL into the institute. NYU appointed two members of the medical school faculty to visit HASL and report on the merits of the proposal. They must have reported favorably to the NYU administration, because things moved very rapidly after their visit. In anticipation of the need to provide space near the medical school campus, centered at 31st Street and First Avenue, a wealthy real estate operator who was chairman of the NYU Medical Center Board of Trustees, took an option on a building owned by the New York Life Insurance Company on the southwest corner of 26th Street and First Avenue. The institute prepared and submitted a formal proposal to the AEC, and for several weeks it appeared that the five AEC commissioners would accept it. However, Charles Dunham had succeeded Bugher as Director of the Division of Biology and Medicine. Dunham took the position that HASL was needed by the AEC as it was—a laboratory staffed by government employees who were fully responsive to the needs of AEC. The building on 26th street was, however, retained by the NYU Medical Center, and became the new home of the School of Dentistry.

Nelson and I were disappointed that the union between the Institute of Industrial Medicine and HASL never took place, but the discussions that took place were not without benefit to NYU, and ultimately had an important effect on my career. Bugher had the idea that when HASL was integrated into the institute there would be a need for a fund that could aid in the transition. He was instrumental in arranging for a $500,000 grant from the Rockefeller Foundation, to be paid at a rate of $50,000 per year for ten years to provide continuity of support for the senior members of the staff. In the 1950s, the annual salary of a full professor was about $17,000, so the Rockefeller grant was substantial by the standards of the time. It served admirably to stimulate the research of the institute.

With my move into full-time administration, I had assumed that my activities as a part-time member of the NYU staff would become less frequent. However, Nelson wanted my advice in the selection of a scientist to lead the program to be established with the Rockefeller funds. A number of excellent candidates were interviewed, but no selection was made for more than one year. In the meantime I was finding that, while I enjoyed my role as an administrator, my real interest was environmental research.

By early 1959 several factors made me think about leaving AEC. Not the least of these was that atomic energy was increasingly of interest to the private sector. Nuclear power was already at hand, and there was a widespread belief that uranium would eventually replace coal and oil as a source of energy. The uses of radioactive isotopes were expanding rapidly in clinical medicine, research, and industry, which resulted in the formation of many consulting and manufacturing companies. The anticipated increase in the demand for uranium resulted in formation of companies for mining, extraction, purification, and reactor fuel fabrication. These developments were creating positions for industrial executives with a background in nuclear science or engineering. Moreover, universities were starting graduate research and teaching programs, and state and federal agencies were looking for personnel to direct their regulatory and monitoring activities. Many opportunities were developing.

Two other factors also began to influence my thinking. I was still in my early forties, and for the past three years had been in the top civil service grade GS-18. I could not increase my income as a government employee, although I could become more influential if I moved to Washington. This I was reluctant to do. I realized that sooner or later a change was necessary, because I couldn't spend the rest of my life as manager of the NYOO, much as I liked the position.

The second factor was that our three sons, Elliott, Michael, and Fred, were now teenagers and would soon be needing college and postgraduate educations which I couldn't possibly provide unless I increased my earnings. This was unquestionably the most compelling reason for considering a change. I had many colleagues in government who were satisfied with their position in all other respects but were forced to leave for the same reason.

Most of the executives who were retiring from the AEC in the late 1950s were taking positions in industry at three times their government salaries. The rapid turnover of AEC executives was having a serious effect on morale and efficiency. It was also the cause of jokes, as when I was at a conference in Chicago in August 1958 and heard one of the Washington staff quip as one of his associates was called to the phone, "If it's my boss be sure to get his name!"

I had often thought that I would like an academic position, but had never really considered the possibility seriously. Nelson had stated many times that he would like me to fill the newly established professorship at NYU. The position interested me very much, but I was afraid that my lack of a doctorate would make me feel uncomfortable in an academic environment.

My departure from AEC began in the early summer of 1959 with a request from Eugene Zuckert, who had previously served as an AEC commissioner and also as assistant secretary of the Air Force, to join him for dinner in New York. I knew that something unusual was afoot, because I had had very little contact with him when he served on the commission. He told me that the board of directors of a newly formed company was dissatisfied with its president and had asked him to inquire about my availability. I knew something about the company and was well aware that it was floundering, despite its blue-ribbon board of directors, which included many of the country's leading nuclear scientists and engineers, two of whom have since received Nobel prizes.

The prospect of becoming the president of a small company with serious financial problems did not appeal to me. However, out of our conversation came the idea that instead of being the company's chief executive officer, I could help by being a consultant to its president. This would require about four days per month. It would not be possible for me to serve as a consultant while I was a government employee, but such an arrangement could augment an academic salary very nicely. I left Zuckert that evening with the understanding that I would retire from the AEC and enter into a consulting relationship with the company if I could make the necessary arrangements at NYU.

The day after my discussion with Zuckert I met with Nelson and asked whether I could accept the NYU appointment on the condition that I would spend about one day a week in private consultation. He replied that this would be possible with a full-time appointment, since members of the faculty were routinely allowed one day per week for consultation and other outside activities. I was amazed at the speed with which things began to move. Our conversation took place just before the Fourth of July holiday, when the medical center faculty were scattered around the world, but Nelson somehow managed to reach the key individuals by telephone and obtain the approval required to offer the appointment. NYU sent me an offer of an appointment as professor of industrial medicine on July 31; I went to Washington promptly and advised my superiors that I would be leaving the AEC in early September.

There were surely many in the Washington headquarters who were glad to see me go. At one time or another I had disagreed with established policies, although more times than not the positions I recommended were the ones adopted. This was true at the very start, when I had opposed the decision to disband HASL. I had been critical of the fact that AEC did not maintain a fallout monitoring network, which I nevertheless established over many objections. I had insisted that intensive monitoring programs be established at intermediate distances from the test sites in Nevada and the Pacific which irritated those on the staff who didn't think it necessary. In short, I had gone over the heads of midlevel headquarters staff members who were blocking actions that were in the national interest and in the interest of safety. I believed that a decentralized method of operation was best for the AEC, and I had on many occasions resisted efforts by headquarters staff who wanted decision making centralized in Washington. Nevertheless, I left many friends in the AEC, and eighteen years later I was awarded a gold medal by the agency. The citation that accompanied the medal read:

For his many contributions to and his meritorious leadership in radiological and environmental health which have provided essential support for atomic energy programs; for his contributions toward establishment and implementation of occupation and environmental threshold limit values for beryllium, the first metal for which this was done; for his services as Director of the AEC's Health Safety Laboratory including the first worldwide fallout monitoring network, and direction of landmark research on the industrial hygiene of the uranium industry. . . .

SEVEN

New York University
Medical Center

B EFORE I JOINED the New York University Medical Center I had taught graduate courses part time for about a dozen years at both NYU and the Columbia School of Public Health. This began in early 1946 when the NYU College of Engineering decided to add a course in industrial hygiene to the graduate curriculum in sanitary engineering. Since there were very few people who knew the subject, I visited the campus to offer my help. Professor Rolf Eliassen offered me a position as adjunct assistant professor, beginning with a one-semester course in the fall of 1946.

At about the same time, Leonard J. Goldwater, a physician specializing in occupational medicine, returned from the navy to become chairman of the Department of Occupational Medicine at the Columbia University School of Public Health. He too invited me to join his faculty on a part-time basis, and for a few years I lectured in short courses from time to time.

Another of the many officers who returned to New York from their military service was Anthony J. Lanza, a sixty-three-year-old physician who had been a vice-president concerned with occupational health at the Metropolitan Life Insurance Company's New York headquarters. During World War II he had served as a colonel in charge of the army's program in environmental health and had established a research laboratory at Fort Knox, Kentucky, where he assembled a remarkably talented group of scientists, physicians, and engineers to undertake research on environmental problems of concern to the army. Among those in the group was Willard Machle, a physician specializing in toxicology, who before the war had been on the staff at the Kettering Laboratory of the University of Cincinnati. The senior engineer at Fort Knox was Theodore Hatch, whose industrial hygiene program at the University of Pennsylvania I had seriously considered joining in 1939. The younger members of the Fort Knox staff included Norton Nelson, a physical chemist, Edward Palmes, a toxicologist, and Roy Albert, a physician who had trained in internal medicine.

Lanza was determined to found an Institute of Industrial Medicine at the New York University-Bellevue School of Medicine, as it was then called. This he did, bringing with him in the process Norton Nelson, Edward Palmes, and Willard Machle, the last in a part-time capacity. Lanza, having learned that I was teaching industrial hygiene in the College of Engineering, asked if I would join his faculty as an adjunct professor. I accepted, not realizing that I would eventually spend more than a quarter of a century as a full professor and laboratory director in the department he was establishing. I had to pay the price of resigning my position at Columbia, because at that time there was a rule that no one could serve on both faculties.

Immediately after the war, Congress established "The GI Bill of Rights," which gave returning veterans an opportunity to continue their education at government expense. Never since have I had classes so swollen as during those postwar years when I was teaching part time. With free education available, the veterans demonstrated a terrific thirst for graduate education. The classes frequently included as many as forty students, far more than the eight to twelve students that I later became accustomed to at NYU, a number which permitted the close classroom interaction that is so important in graduate education.

I joined the institute's full-time staff in September 1959 as professor of industrial medicine (environmental medicine after 1962) and director of the Environmental Radiation Laboratory. By that time Norton Nelson had succeeded Lanza as director of the institute, a position he would hold with distinction for twenty-five years. Upon his retirement, Arthur C. Upton moved to that position from the directorship of the National Cancer Institute. I count it a privilege to have served under these two great scientists.

As a full-time professor, I erroneously thought that finally I would be able to earn my Ph.D. While a part-time teacher, I had already taught nine graduate credits, and I thought I could surely receive credit for courses that I had taught. I assumed I could eventually complete the remaining course requirements and since I would be involved in an active research program, it should be a simple matter to complete a dissertation. What I didn't know until after I joined the NYU faculty was that, as in most universities, no one can be awarded a degree by a faculty of which he is a member. However, unknown to me (I have since suspected John Bugher), a movement was afoot to award me an honorary doctorate. One day I received a letter from Fairleigh Dickinson University advising me that its board of trustees had voted to award me the Doctor of Science (*Honoris Causa*). The letter expressed the hope that I would accept the degree, which I did with alacrity. I would have much preferred to have "earned" the doctorate, but since circumstances did not give me the opportunity to complete the doctoral requirements in the usual way, I was delighted and gratified to receive the honorary degree. Several years later I was similarly honored by the Catholic University of Rio de Janeiro.

I hoped that the initial focus of the laboratory would be to improve our understanding of the chemical, physical, and biological processes by which radioactive materials are transported through the environment, from the time they are introduced as contaminants until they are inhaled or ingested by humans. This would be our focus, not only for the research program, but also for the instruction of graduate students. I also hoped that we could collaborate with state and city health departments as well as other departments of the medical school in studies of the effects of medical or occupational radiation exposure on people. This proved to be as useful an orientation for chemical as for radioactive contaminants. After a few years our research and teaching extended into the field of chemical toxicology and the name of the laboratory was changed to the Laboratory for Environmental Studies.

An important objective of the teaching program was flexibility in providing Ph.D. training to students with varied educational backgrounds. I did not want to establish strict requirements for prerequisites that would exclude students who were worthy in all respects except for deficiencies in undergraduate courses. With such strict requirements, a student who had a degree in biology, but was deficient in mathematics or physics, was required to complete the requirements without credit at the same time he was taking graduate courses with credit. This often lengthened the time required to complete the course requirements for the Ph.D., especially since the student might be required to defer certain courses for which the prerequisites had not been completed.

I attached great importance to flexibility because of my experience with academic compartmentalization in the College of Engineering. There I had been teaching industrial hygiene to students who were preparing for graduate degrees in sanitary engineering, a sub-specialty of civil engineering. Civil engineers were traditionally concerned with the construction and operation of water supply and waste water treatment systems, so it was understandable that they should be interested in environmental health generally. However, civil engineering graduates were often not well prepared to study industrial hygiene or to use it in their work. The course would have been of equal or greater benefit to the chemical engineering students who would later design and operate potentially hazardous industrial processes, but it wasn't available to them. The departmental requirements also specified a bachelor's degree in civil engineering as a prerequisite for graduate studies in sanitary engineering, which further narrowed eligibility for the course.

Nelson imposed an important restriction on our program that proved to be very wise. A danger with interdisciplinary teaching programs is that the courses can be so superficial that the student is not subjected to the discipline required to master a subject in depth. The knowledge gained is broad but shallow. To overcome this problem, Nelson insisted that the student complete the requirements of one of the basic science or engineering departments. This was a necessity for another reason: By the terms of the charter of the medical school from the Board of Regents, our department could award degrees only to students who had already earned an M.D. To broaden the base of students, it was thus necessary for us to develop collaborative programs with departments in the Colleges of Engineering or Arts and Sciences, into which the students were required to matriculate. They were required to complete the requirements of that department and college, but undertook their research and prepared their dissertations while in residence in our department. The students were also required to take certain specified courses offered by our department. The degrees—M.S. or Ph.D.—were thus actually awarded by the department with which we were collaborating.

This arrangement worked well for many years but required that the student take many more courses than were normally required for Ph.D. preparation. Since the fellowships for student support were modest, and the course requirements tended to keep the candidates in residence longer than was normal, we made every effort to employ students as research assistants at salaries appropriate for their level of experience. This attracted excellent students who had experience in the field and helped the Environmental Radioactivity Laboratory get off to a quick start. For example, one of our early students was Henry Petrow, who came to us with a master's degree in chemistry from Harvard and about ten years of excellent research experience in a commercial laboratory. He was not only a student, but also set up and supervised our radiochemical laboratory.

The fall of 1959 proved to be an excellent time in which to begin a training program. Two months after my arrival I was visited by Russell Morgan, a prominent Johns Hopkins radiologist who had been appointed chairman of the newly formed U.S. Public Health Service National Committee on Radiation (NACOR). He told me that the Public Health Service was in the early stages of developing a program to support graduate training in radiological health and encouraged me to prepare a proposal for such support. This was of particular importance to newly formed centers such as ours. Until then such training centers were AEC supported at a few universities such as Rochester and Vanderbilt, where the emphasis was on training and research in the basic radiological sciences. The Public Health Service, however, was interested in training practitioners who could become directors of the radiation protection programs needed in the states and teach radiation protection in the colleges. Moreover, although public concern about radiation hazards tended to focus on the military applications of atomic energy, the largest man-made contribution to the dose received by the public was from X-rays used in medical practice. The Public Health Service considered that it was its responsibility to advise the states on minimizing the exposure. As a result of Morgan's visit, our laboratory applied for and received a training grant that for about ten years funded much of the support we provided to our graduate students.

The laboratory was established at a time when "whole-body counters" were becoming popular as a research tool. They were basically small rooms with walls constructed from six inches or more of steel, and equipped with sensitive radiation detection equipment. They were called whole-body counters because they could measure the amount of gamma-radiation-emitting substances in the human body. At the time it was thought such counters would be useful in clinical research, but the six or eight of them constructed in the U.S. were located at National Laboratories and university physics departments rather than at medical research centers. HASL had located components from battleship gun turrets that would make two small but otherwise excellent shielded rooms, and arranged to have them shipped to New York, where one was installed at HASL and the other in the basement of the Medical Sciences Building of the NYU Medical Center. The Atomic Energy Commission reasoned that a whole-body counter at the NYU Medical Center would provide a useful research tool for investigators in the New York City area. The whole-body counter served as the centerpiece of our laboratory until we moved to Sterling Forest in 1962. Thereafter, the unit at the medical center was taken over by others at the institute, where it remains an important research tool today.[1]

The only financial resource at the founding of the Laboratory for Radiation Studies was the Rockefeller Foundation grant, and it was all we required. The first staff member after myself was a secretary, soon followed by Gerard Laurer, an electrical engineer who had just received his master's degree in radiological health from the University of Rochester. As a research assistant and graduate student, he was immediately placed in charge of the whole-body counter, as well as any other radiation counting equipment we acquired. Roy Albert, who had been with the group at Fort Knox and then worked on the HASL staff as medical officer, also joined us, as did Bernard Pasternack, who had recently received his Ph.D. in biostatistics at the University of North Carolina School of Public Health. Hanson Blatz, who had left HASL to become head of the Radiation Protection Branch in the City of New York Health Department, was on our part-time staff and taught the first courses in what others called health physics, but we called radiation hygiene. With this nucleus we started in the fall of 1959; from it evolved a teaching and research unit that during the next quarter century attracted students from many states and foreign countries. Before long, Pasternack split off to form his own laboratory in biostatistical and epidemiological research, as did Albert, who formed a laboratory for experimental medicine.

From late 1959 until 1963, my laboratory was in the basement of the Medical Sciences Building, which was built on filled land on the bank of the East River. Sometimes at high tide, water carrying odoriferous hydrogen sulfide oozed through the floor. Although the space left much to be desired, I was appreciative of every square inch because I knew how difficult it was to find space at the Medical Center. Many scientists say that a crowded laboratory brings out the best in young investigators. That was certainly true during the three years we spent in our basement, where the work done by the graduate students was unsurpassed both in quality and volume compared to any other time in my experience. In a crowded laboratory everyone is visible, and new ideas are quickly communicated and discussed.

We left our basement laboratory in 1963 when the institute began a move to new quarters in Sterling Forest, near Tuxedo, New York. The forest was a thirty thousand-acre tract of craggy hills, lakes, and meadows that had long been part of the estate of the Harriman family, but was about to be developed as a community of research laboratories and homes. New York University intended to consolidate its research facilities on a thousand-acre tract donated to it for that purpose; our institute was the first to move to what was to be a new campus. Irma and I sold our home on Long Island and built a new one on the west shore of the seven hundred-acre, crystal clear Sterling Lake, where we were to spend the next twenty-three years. The one-acre tract on which we built the house was leased to us for ninety-nine years by the property owner, the Sterling Forest Corporation.

The hemlock and hardwood forest in which we lived had been essentially undisturbed for generations, and was extraordinarily remote, considering that we were only about forty miles from New York City. Deposits of iron ore were discovered throughout the area in colonial times, with the result that the remains of old mines and furnaces long since encroached upon by the fern, laurel, and hemlock surrounded us. One rich vein of hematite, located 800 feet below Sterling Lake, had been entered for more than 150 years through a shaft located a quarter mile south of our home, with the remains of the furnace where the ore had been smelted still standing just beyond. The forest was laced with the beds of old narrow gauge railroads over which mule-drawn carts once hauled charcoal, wood, limestone, ore, and iron.

The lone caretaker of Sterling Forest was Tom Whitmore, about sixty years old, who lived with his wife Olive in the cottage in which he was born on nearby Longmeadow (once Longswamp) Road. I learned about the remotest areas of the forest during many jeep trips with him, when Irma and I were often accompanied by dogs and grandchildren. One of his main responsibilities was to control poachers, which was a dangerous responsibility because many were descended from the original settlers and recognized none of the restrictions imposed by the owners of the Sterling Forest tract. About ten years before we moved to the forest, a game warden had disappeared under circumstances that suggested he had been murdered by poachers who disposed of his body down one of the many flooded mine shafts.

Wild game was plentiful. White-tailed deer, red and gray fox, otter, beaver, and raccoon appeared almost daily, and three varieties of trout as well as wide-mouth bass were in the clear waters of the lake. The lake was also the home of many varieties of ducks and geese, in abundance. During the first of the harsh winters, we maintained a rack of alfalfa for the deer, which yarded down outside our living room window by the dozen. We were cautioned by my ecologist friends that this was a poor practice because the vigor of the species could only be maintained by the culling that took place naturally each winter. We eventually gave up the practice of feeding the deer during the winter months, but for other reasons. They became a nuisance and showed no appreciation for the hospitality we offered them. The deer enjoyed the alfalfa that was intended for them, but then nibbled on our plantings, including the many mountain laurel and rhododendrons I had transplanted from the forest to our property.

We learned a lot about nature as we watched the deer from our living room windows. Although they are gentle in appearance, we found they were not always kind to each other. In the middle of winter when we were hosts to a herd of a dozen or more, a lost orphaned fawn would sometimes emerge from the forest and approach the rack of alfalfa, only to be attacked by the sharp hooves of the herd members. I soon learned to help the orphans by placing a small amount of the feed at the edge of the forest, some distance from the hay rack.

During the winter months feral dogs yelped through the night as they tracked the deer through the forest. Occasionally they drove a deer onto the frozen lake, where the deer's legs became splayed and it became immobilized on the ice. The dogs then attacked it, not for food, but seemingly for the love of a kill. After a few bites, the dogs left the mortally injured animal to die on the ice for the benefit of the foxes, crows, and turkey vultures that fed off the carcasses until the ice left the lake in the spring.

We soon assembled our family flotilla, consisting of a canoe, rowboat, and sailboat, which gave us much pleasure over the years. Power boats (other than electric motors) were forbidden. By living seventy-five feet from the late (the minimum allowed distance), we enjoyed boating in the warm weather, and ice skating in winter. We enjoyed many days of snow shoeing through the woods, where we studied the habits of the wildlife by following their tracks through the snow.

The laboratory was located exactly one mile around the lake from our home, and most days I walked back and forth, often more than once. At the end of the day I would usually telephone Irma who would meet me half way, at the foot of the lake, accompanied by one or more of our dogs. For two people who loved the outdoors, it was a marvelous way to live for nearly a quarter of a century.

Our institute was to have been in the vanguard of many New York University graduate research facilities that were to be relocated to the forest. Unhappily, NYU, the largest private university in the country, encountered severe financial problems shortly after our arrival, and the plan was scrapped, leaving us with only one other unit of the university in the forest, the Laboratory for Experimental Medicine in Primates.

At first the institute was in an old schoolhouse built after the First World War for the children of the miners. It had been used for a few years as a staging area by Union Carbide, which was building a major facility a few miles to the north. Since Carbide was ready to move into permanent quarters at the time we were ready to send our first laboratory units to Sterling Forest, we purchased the building and renovated it. We eventually added two additional buildings, and named the complex the Anthony J. Lanza Laboratory of the Institute of Environmental Medicine.

The grant from the Rockefeller Foundation was intended to establish a teaching and research laboratory in the radiological sciences and was to have been distributed over a ten-year period. After only two years we found that the team we assembled was attracting research and teaching support to such an extent that the Rockefeller funds were no longer needed. Nelson obtained permission from the foundation to apply the balance of grant to the cost of constructing the Lanza Laboratory.

Sometimes a combination of circumstances accidentally has an effect on future events. One of the problems that arose out of our move to Sterling Forest was that we had to leave our whole-body counter behind. It was a massive structure that could not be moved like other laboratory equipment. Even in the mid-1960s, a whole-body counting room cost in excess of 100,000 dollars. At first it wasn't obvious how we could finance one for Sterling Forest. Unfortunately the matter could not be deferred for very long because it could best be put in place before the building under construction was closed in. Otherwise it would be necessary to tear down an exterior wall to permit placement of the large pieces of steel.

In January 1966 the problem was solved by an accident, both literally and figuratively, when a U.S. Air Force B-52 bomber crashed in Spain, with the loss of four thermonuclear weapons. There were no nuclear reactions, but the chemical explosives in two of the bombs exploded, scattering plutonium over the coast. There were no injuries, but there was concern that some of the inhabitants might have been exposed to plutonium dust.

It occurred to me that since accidents of this kind were likely to happen again, the Air Force would be well advised to develop a transportable system for scanning the bodies (particularly the lungs) of persons suspected of having been exposed to plutonium dust. The problem with existing instruments was that plutonium emits alpha particles which were not detected by gamma radiation equipment of the type we had used for measuring the radioactive iodine content of the human thyroid. However, Pu-239 also emits a very soft X-ray. I asked Gerry Laurer if it was possible to detect the small fraction of that emission that leaves the body. He proposed a system of scintillation counting that involved a thin crystal of cesium iodide and a thick crystal of sodium iodide, optically coupled to a photomultiplier, from which the light pulses from the two crystals could be differentiated electronically. Laurer developed the system as part of his doctoral dissertation; it is now the standard method of measuring the body burdens of radioactive elements that emit only soft penetrating radiation.[2] However promising Laurer's proposed instrument was, a whole-body counter was essential to its development. We simply could not proceed without one, because it was necessary to test and calibrate the instrument in a radiation-free environment.

When I explained to the Department of Defense that we could build an instrument that could be taken anywhere in the world to measure the amount of plutonium in human lungs, there was immediate support for its development. When I added that we could not build and properly test the instrument without access to a whole-body counter at the Sterling Forest Laboratory, the department added about a hundred thousand dollars to the cost of developing the plutonium detector, which itself cost only roughly half as much. Of course, under the terms of the grant, the shielded room belonged to the government, but there was no practical way to remove it from where it was installed.

After about a year, Laurer's instrument was delivered to DASA, the

Defense Atomic Support Agency of the Department of Defense, where the officials were very impressed by his ingenuity. To my knowledge a need for its use has not since arisen. In any case, we had a shielded room we would likely not have had, but for the Palomares accident. For us it was a fortunate accident indeed. A generation of graduate students who have helped develop the art of measuring radioactivity in the human body has used the room.

By the time we moved to Sterling Forest, the research interests of my group had begun to diversify. This had already been indicated by the change of name to the Laboratory for Environmental Studies. In the new facility we had ample space and basic laboratory furniture, but it was necessary to find outside research support to provide special equipment and apparatus.

Research within the academic system in the United States is based on the "principal investigator." The academic institution functions as the housekeeper. Where the institution is involved in both teaching and research, as is the case in most graduate schools, the school must oversee teaching activities to assure that they are properly integrated and of the required quality. No such oversight, integration, coordination, or quality assurance is provided for the research programs, however. Yet, measured by the amount of money involved and the size of the staff required, research overwhelms teaching in the large academic centers. The research performed is usually up to those scientists that qualify as principal investigators by their ability to obtain funds, attract students, and publish their research findings in respected scientific journals. An investigator accumulates "credits" through his publications in the scientific literature, first as a colleague of a scientist with considerably more experience and eventually, if he is successful, as a "principal."

The principal investigator and his associates depend on government agencies, foundations, or other sources of private sector support. In preparing research applications, the investigator demonstrates his mastery of the subject. Prior literature must be reviewed, and the proposal for a new or continued investigation must be clearly described. The application must also describe the results of the applicant's previous studies, and cite his or her prior publications. The proposal will then be reviewed by peers of the investigator who are specialists with established reputations in the same or allied fields, and who submit their critiques to the sponsoring agency. There are usually more requests than can be accommodated by the available funds. Unfortunately, many grant applications fail to obtain support, not because of poor quality, but because they fail the test of relevance. An agency will not accept a proposal if it is perceived to be outside the bounds of its interest. This can be a serious problem to investigators who work in the basic sciences, because the subject of their interest may not yet have been included among the programmatic interests of the agencies.

Fortunately, my research interests and those of my colleagues were in applied science. Our research was of immediate practical importance, whereas the basic scientists were at the frontiers of knowledge and often did not know or care whether the results of their research would be of practical value beyond adding to the total of human knowledge. I would like to believe that our group attracted ample support because of the excellence of the research we conducted. However, I also realize that we became a popular laboratory among the supporting agencies because we enjoyed finding solutions to some of the many practical problems they faced.

Microwave Workers

As the result of Norton Nelson's involvement in the Armed Forces Epidemiological Board, it was decided to determine whether opacification of the lens of the eye could result from exposure to microwaves. An Air Force physician who had examined a small group of radar workers had reported that minor lens imperfections were present more frequently in the eyes of the workers than among the controls. Although the imperfections were so minor that they were not considered of clinical significance, it was possible that such changes were progressive, in which case cataracts might be developing with increased frequency among microwave workers and might also occur at an earlier age. Experiments with laboratory animals had shown that cataracts could be produced at high levels of microwave exposure. Moreover, cataracts had been reported among microwave workers, although it was not known whether the cases were more frequent than in an unexposed population of similar age.

Minute changes occur in the lens of the eye that are not of themselves considered to be of clinical significance. Over a period of many years, some of these changes may be related to the process of opacification, but little is known about them. Ophthalmologists do not ordinarily record their presence, and at the time of our study (1960) a standard method of classifying them did not exist.

The way in which this project was conducted illustrates some of the features of multidisciplinary investigations. When it was decided that the study was needed, I agreed to serve as principle investigator. The Air Force would provide financial support. The Department of Ophthalmology was willing to collaborate with us and assigned a clinician as coinvestigator. Bernard Pasternack was our biostatistician and a graduate student would serve as research assistant. Stephen Cleary, who had just completed his master's degree at Rochester, was selected.

In addition to organizing the research team, I developed the research plan. This required that I become familiar with the various kinds of microwave generators and the ways in which workers were exposed to their emissions. It was then necessary to develop a method for assigning estimates of the severity of microwave exposure. This would have been a simple matter if systematic measurements of exposure had been made, but this had not been done. I thus decided to base the severity of exposure on occupational histories taken by myself and Steve Cleary. We assigned weights to such factors as the power levels of the equipment used, distance from the antenna, number of months of employment, and even whether, when working near the generators, they had sensed the heat being produced in their bodies because they were in a microwave field.[3]

The next step was to select the population to be studied. For this, we wanted men with relatively long periods of employment who had jobs in which it would be expected that microwave exposure was occurring. We finally agreed on a study population of more than 2000 men, including those exposed and age-matched controls, located at military bases and industrial facilities throughout the United States, and at the Air Force Base above the Arctic Circle in Thule, Greenland.

Now that we had identified the study population and had developed a method for scoring the level of microwave exposure, it was necessary to decide on the kind of eye examination to be conducted and a system of scoring the severity of the observed lenticular imperfections. One problem was that there had been so little interest in the changes in the lens, that the various types of imperfections had not been classified. After many hours of discussion with our ophthalmologist, it was decided that the changes seen could be divided into five types, and that for purposes of scoring, each type would be described according to a scale of 1 to 3, in which grade 1 would indicate only minimal changes. The ophthalmologist selected the method of examination. In this he was assisted by the fact that thousands of the survivors of the atomic bombings in Japan had been examined with a photographic slit lamp that provided a permanent record of the condition of the lens. It was decided that the methods used in Japan were ideal, and the equipment developed for that program was loaned to our project. The next step was conducting the eye examinations in the field. A strict requirement was that the ophthalmologist must not be aware of the severity of microwave exposure sustained by the individual being examined, or whether the individual was a microwave worker or one of the control subjects. This avoided the possibility of biasing the results of the examination, consciously or otherwise.

When the examinations were completed, Cleary and Pasternack analyzed the relationships between the reported lens changes and the levels of exposure to microwaves and found that there was a statistically significant relationship between the estimated levels of exposure and the frequency of occurrence of the minor lenticular imperfections. It was also demonstrated that the number of imperfections increased with age, even among unexposed individuals, but that the increase occurred more rapidly among the microwave workers. Thus, at any given age, the lenses of the microwave workers had the appearance of those in unexposed men who were a few years older. The clinical significance of this finding was not evaluated as part of this study.

Many years later I attempted to obtain support for a reexamination of the exposed workers to determine if the lens changes had progressed. Unfortunately, as often happens, interest had waned and the required support was not available.

In 1969 I was appointed to the Electromagnetic Radiation Management Advisory Council which had been established by the Office of Telecommunications Policy (OTP) in the White House. The basic purpose of the council was to coordinate the research programs sponsored by various government agencies, but we were also consulted from time to time when policy guidance was needed by the executive branch.

One of the first problems after I joined the council arose because the Soviet government was found to be deliberately spraying the U.S. embassy in Moscow with microwaves in the radar frequencies. The reasons for this must have been known to counterintelligence officials, but was never explained to the council. However, our members received information on the frequency spectra and field strengths involved. It was apparent that whatever purpose the Soviets had, the field strengths were so low as to be insignificant with respect to biological effects. However, the Soviet practice was given considerable publicity, as a result of which apprehension developed within the embassy staff. This was understandable, since our government had filed protests with the Soviets on the grounds that the irradiation might be endangering the health of the staff. This was ill-advised, because the field strengths were no greater than those associated with operation of automobile two-way radios. Because of the fears of the embassy staff, however, the ERMAC members decided it would be advisable to issue a statement that the levels of exposure were far too low to be of concern. A brief statement to that effect was drafted and approved by all council members but one, a physician who was a consultant to the State Department. He suggested to the chairman, an OTA staff member, that the matter should not be pursued further, whereupon the subject was dropped. It became clear to me that our government was more interested in embarrassing the Soviets than in providing much needed and justifiable assurance to the embassy staff.

My interest in the non-ionizing radiations continued for many years. In the mid-1970s, I encouraged a postdoctoral student, Asher Shepard, to review the literature on the effects of the electromagnetic and electric fields of extremely low frequency (below 100 Hertz), which resulted in a book that we published in 1977.[4] I also began research aimed at determining whether the biological effects produced by the radar frequencies were due to mechanisms other than those explained by the thermal effects of exposure. In collaboration with Sidney Belman, a biochemist on the faculty of our Institute, and James Rabinowitz, a postdoctoral student, we designed an experiment that demonstrated that non-thermal effects did indeed occur.[5] I believe it was the first time this was accomplished.

Radiation Exposure from Medical X-rays

By 1960 it was already known that the medical uses of X-rays were a major source of radiation exposure, but no quantitative information was available. Hanson Blatz, who held appointments on our staff as well as in the New York City Department of Health, recommended that a survey be conducted to obtain information about the doses received by patients from the use of X-rays in physicians' offices. This was a pioneering effort that would not have been possible without the close association we enjoyed with the Health Department. The first problem was that no one knew how many X-ray machines there were in New York City, or where they were located. Since it was anticipated that the Health Department might want to inspect the machines on a routine basis in the future, the department decided to establish a registration system. Our research was absolutely dependent on the data-base from that registration and inspection program.

When we were satisfied that most of the machines were registered, a sample was selected based on the type of X-ray machine, the procedures for which it was used, and other factors. To assure their cooperation, there were conferences with the physicians selected to be part of the program. Forms sent to the physician's assistants requested that certain information be recorded each time the machine was used. This included the age and sex of the patient, the type of examination, the tube current and voltage, and the procedure involved.

At the time the study was planned there was more concern about the genetic effects of radiation than its ability to cause cancer, which is the reverse of our understanding of radiation effects today. To evaluate the genetic effect, one is more concerned with the dose received by the population than that received by the individual. The term "genetically significant dose" had been conceived to describe the average dose received by individuals in the population, weighted for factors that determine whether the exposure could have a genetic effect on the population. The dose to the testicles and ovaries must be estimated, since these are the organs in which the genetic mutations take place. Gonadal exposures to persons past the age of childbearing can be ignored, and the "genetically significant dose" from exposure of persons in the age of procreation are influenced by the number of children expected.

An enormous amount of data was accumulated in the survey. We depended on Bernard Pasternack to assist with the biometrics. Mortimer Heller wrote his doctor dissertation on the data obtained. This study demonstrated the prevalence of poor practices among the physicians and their technicians, many of whom did not understand the importance of minimizing the exposures to X-rays. It was known to be important to columnate the X-ray beam so that only portions of the body being X-rayed would be exposed. Despite this, the gonads were within the primary beam in fifty percent of the chest X-rays included in the survey.[6]

Resumption of Nuclear Testing in the Atmosphere

In 1958 the nuclear powers had agreed to a moratorium on atmospheric weapons testing that was broken unilaterally by the Soviet Union in August 1961. About a week later, our laboratory began to find iodine 131 in milk being sold in New York City. One night Jerry Laurer and I were working late in the laboratory. Because I knew that he consumed more than a quart of milk per day, I suggested that he step into the whole-body counter and place the front of his neck against our eight-inch sodium iodide scintillating crystal to see if his thyroid had accumulated the radioactive iodine from the Soviet bomb. It took only a few minutes to demonstrate its presence in a small but readily discernible amount. I then followed the procedure on myself, but no radioactive iodine was detected in my thyroid because I consumed hardly any milk. Thus began the resumption of the fallout studies I had begun at HASL.

Iodine 131, which has a half-life of eight days, is produced copiously in nuclear fission. It is readily detected by gamma radiation spectrometry, but that analytical technique was not generally available until the late 1950s and little was learned about the behavior of iodine 131 in the environment during the tests conducted prior to the 1958 moratorium. Iodine 131 enters the body in several ways. When present in the atmosphere it is absorbed by inhalation, but this is a minor source compared to dairy products, because the radioactive iodine settles on grass from which it is absorbed by grazing cows. Of course, the cows also inhale the I-131, but the cow grazes such a large area that the inhaled radioiodine is insignificant compared to what is ingested by grazing on contaminated grass. The I-131, like stable iodine, concentrates in the thyroid. Thus, children who consume milk tend to receive the highest doses from an environment contaminated by I-131: not only do they consume more milk per unit of body weight, but their thyroids are smaller, and the dose delivered by the radioiodine is more concentrated.[7]

One of our first objectives was to study the relationship between the concentration of radioiodine in milk, and the amount in the thyroids of children. To start with, we needed crystals smaller than the eight-inch one we had. I telephoned James Terrill, Director of the Bureau of Radiological Health of the U.S. Public Health Service and explained our need for equipment that would enable us to undertake the study. He agreed that we should purchase the equipment immediately, and that he would reimburse us through the training grant we had recently received from his division. Fortunately, the equipment was readily available, so we were able to begin our measurements in a few days. We made measurements on people of all ages, but our most useful source of subjects was the outpatient clinic of the Pediatrics Department, where we found the mothers very cooperative, since they were amused by coming to our laboratory and entering our whole body counter with their children. A lollypop or two was also part of the reward we offered to the hundreds of children measured in this way during the next two years. We learned a great deal about how dairy products become contaminated by radioiodine and the mechanisms by which the radioiodine enters the thyroid.

One day a mother came to the laboratory with two small children who had been to the outpatient clinic of the pediatrics department. The mother told us that both children drank about the same amount of milk each day, but our measurements showed that only one of them had the expected amount of thyroidal I-131. The other child had almost none. We learned further from the mother that they had visited the clinic because one of the children (the one who had accumulated almost no I-131) was asthmatic, and was being medicated with Lugol's solution, which contains a small amount of stable (ordinary) iodine. The child's blood therefore contained more than the usual concentration of stable iodine, which had the effect of diluting the radioactive iodine and reducing the probability that an atom of radioactive iodine would be absorbed by the thyroid. We had known that relatively large doses of stable iodide could block the thyroid from absorbing radioactive iodine administered clinically, but there had been no prior indications that the thyroid could be blocked from absorbing I-131 by the small amounts of stable iodine contained in Lugol's solution.

As a result of this experience I enlisted the assistance of Manfred Blum, a thyroid specialist with an interest in radiation medicine. We carried out an experiment with volunteers from the laboratory staff which demonstrated that administration of as little as 250 milligrams of sodium iodide resulted in almost complete protection of a person exposed to I-131.[8] The information we obtained has proved useful in protecting nuclear reactor workers who may be accidentally exposed. A quarter of a century later, small doses of sodium iodide protected the population living near the damaged reactor at Chernobyl in the USSR.

Nowadays it is much more difficult to do studies that involve human volunteers. Twenty-five years ago it was possible to recruit people of all ages for noninvasive studies such as ours, in which there was literally no physical contact with the subjects once they were seated in the chair in our shielded room. Investigators are now required to obtain signed "informed consent" forms from the subjects. The complexity of these forms is a deterrent to even the simplest kind of human research.

One of the first of our foreign postdoctoral students, Yoshio Mochizuki, came to us from the medical faculty of Kanasawa University in Japan shortly after the Soviets resumed testing. He worked closely with Laurer, and calibrated the thyroid measurements we were making in an efficient, but unusual way. Calibration was necessary so that we could relate the measurements of gamma radiation emitted from the necks of our subjects to the quantity of I-131 present in the thyroid. This would normally be done by constructing models (called "phantoms") of the thyroid gland and neck, into which known amounts of I-131 would be inserted. Since the relationship we were seeking would be influenced by the size of the person, phantoms having a range of sizes would be required.

In New York City, all accidental deaths, as well as many other kinds of sudden deaths, must be referred to the medical examiner, a physician who is appointed by the mayor. About twenty autopsies were performed each day at the famous Bellevue morgue, which is adjacent to the NYU Medical Center. Over the years the facilities of the Medical Examiner have been a valuable resource for investigators who needed access to human tissues. For our calibration, cadavers for which autopsies were required were taken to our laboratory early each morning for a few minutes and the neck radiation measurements were made by a technique similar to that used in our studies of living persons. The cadavers were immediately returned to the morgue, where the autopsies were usually performed by midmorning, at which time Mochizuki weighed the thyroid glands and measured their radioactivity. His work not only made it possible to calibrate the procedure we were using, but also provided a unique set of data on the relationship between thyroid weight and age. This data was necessary to compute the thyroid dose, especially for children. The radioiodine studies we conducted resulted in a better understanding of the mechanisms by which human exposure occurs at great distances from nuclear explosions and emphasized the importance of fresh dairy products as the principle vector for human absorption.

As part of the fallout studies we also arranged for samples of lung tissue obtained at autopsy. These were examined for the amounts of radioactive fallout that had been inhaled. MacDonald E. Wrenn, one of our graduate students who later became a member of our faculty, undertook these studies as part of his doctoral dissertation. His study made it possible to estimate the dose of radioactive dust received by the lungs of New York City residents, and also to verify the mathematical models by which the lung dose is calculated from measurements of the concentration of radioactive dust in the atmosphere.[9]

At the time of our fallout studies, much of the research being conducted in a number of countries concerned the geophysical distribution of the strontium 90 produced in test explosions. These studies suggested that deposition on the oceans was greater than on land, even allowing for the greater surface area of the oceans. One hypothesis was that this resulted from the scavenging action of ocean spray in the first few meters above the surface. If so, this was an important mechanism for removing all dusts from the lower atmosphere.

Peter Freudenthal, a graduate student, tested this hypothesis with great ingenuity in experiments he conducted at sea. He demonstrated that whatever scavenging did take place was not a significant deposition process, and that the apparent excess of ocean deposition was a result of uncertainties in the existing data.[10]

Natural Radioactivity in Brazil

In 1961, I began collaborations with Brazilian scientists that have lasted to the present. The research concerned areas in Brazil in which the people are exposed to abnormally high levels of natural radioactivity. I had heard vaguely that such areas existed in Brazil, but I knew next to nothing about them until the establishment of UNSCEAR in 1955. The Brazilian representatives were Professor Carlos Chagas, a physician who was director of the Institute of Biophysics at the Federal University in Rio de Janeiro, and Father Francisco Roser, a Jesuit priest of Austrian origin, who was chairman of the physics department of the Catholic University of Rio de Janeiro. Roser was assisted in Rio by Father Cullen, a physicist from Fordham University who was spending part of the year in Rio. The two priests had loaded their laboratory equipment on a jeep and during the previous two years had travelled through much of Brazil mapping the areas of abnormally high radioactivity.

One day in 1956 Roser told me about their work and I expressed interest in having some samples of the radioactive rocks, sands, and plants for laboratory study. Several months later I received a call from airport inspectors who had found some radioactive baggage that belonged to a priest who said he knew me. This was during the cold war when the U.S. government was so nervous about atom bombs being smuggled into the country that all ports were equipped with sensitive radiation detectors and the holds of incoming ships, aircraft, and passengers' baggage were monitored. The measurements were being made secretly to avoid alarming the public; the only good that came from the program was that the customs inspectors found many wristwatches with radioactive luminous dials that would otherwise have been smuggled into the country. I was still with the AEC when I received the call, and immediately surmised that the priest was Roser, which turned out to be the case. He came to my office the next morning with an amazing collection of material. He had jars of monazite sand from beaches north of Rio where the radiation levels were one hundred times normal, and where the external radiation levels in the nearby towns were nearly ten times normal. He had plants from the state of Minas Gerais which were so radioactive that they could be autoradiographed with ordinary X-ray film. All of this was very exciting to me. Shortly thereafter I was invited to Brazil as a consultant to the Pan American Health Organization to see those interesting places and to advise the scientists at both Federal University and Catholic University (which is everywhere called PUC, short for Pontificia Universidade Catolica) as to the kind of research that should be undertaken. Shortly after I moved to NYU I made the first of about thirty trips to that wonderful country. Strong bonds of friendship and respect developed between the staff of our laboratory and the students and faculties of the two Brazilian universities. We have published many joint research papers, and have also exchanged graduate students.

There are two major radioactive anomalies in Brazil, the monazite sand beaches, and the Morro do Ferro (mountain of iron).[11] Monazite is a rare-earth mineral that contains appreciable amounts of a radioactive element, thorium, and occurs in sands in several parts of the world. Although monazite itself is honey colored, the sands in which it occurs usually contain ilmenite, a mineral of iron that imparts a black color to them. The "black sands" of Brazil stretch for great distances along the Atlantic coast, starting about 150 kilometers north of Rio, in the state of Espírito Santo.

The second major anomaly, the Morro do Ferro, is located in the center of a large caldera in the state of Minas Gerais, near the city of Poços de Caldas. The "mountain of iron" is really only a hill about 250 meters high, near the summit of which is located an ore body that contains massive amounts of rare earths and thorium. Although the thorium is very insoluble, and is therefore not absorbed by the plants that grow in the radioactive soil, the much more soluble radioactive decay product, Ra-228, is absorbed to such an extent that some species of plants can be autoradiographed by placing them on X-ray film for a few weeks. The radiation levels one meter above the ground at the top of the hill are about three milliroentgen per hour, which is more than 300 times the normal level.

A curious thing about the two types of anomalies is that they are exploited for the supposed benefits of the radioactivity to health. The city of Guarapari in the center of the monazite region is a tourist center in which there is a Hotel Radium, and Hotel Thorium. People go to the beaches to sit on the black sand and even rub it on their bodies. Local physicians counsel the tourists on the use of the sands to cure disease, and some tourists take pouches of the sand home where it is used for poultices to alleviate the pain of arthritic joints. The radiation level over the black sands is about two hundred times normal, and in the city of Guarapari it averages about six times normal. The dose received by the local population is higher than normal, but not high enough to produce measurable increases in the incidence of cancer or genetic mutations, although a study conducted in the mid-1960's did find an excess of chromosomal mutations in the white blood cells of the residents.[12]

Cancer and genetic mutations may occur in Guarapari at a higher than normal frequency, but the population of about six thousand is not large enough for the small expected increase to be detectable, even with the best of epidemiological techniques. However, on the southwest coast of India, in the state of Kerala, here exists a much more densely populated monazite region where it might be possible to conduct meaningful epidemiological studies. For more than thirty years, many scientists, including myself, have been urging the Indian health authorities to conduct such studies, but they have not been undertaken. I visited Kerala at the invitation of the Indian government in 1980. After several conferences with their scientists I left in the belief that they would initiate the much needed study, but it has not yet begun. This is a pity, because the Indian monazite region may provide one of the few opportunities to study the human effects of radiation at such low levels.

The Morro do Ferro, being a small hill, is not inhabited, although small farms are located nearby. During our first visits we concluded that the doses received by indigenous burrowing rodents must be very high since thoron, which is a form of the radioactive gas radon, is produced by thorium and would be expected to exist in high concentrations in the soil gases. Robert Drew, one of our graduate students, went to Brazil in 1963 for eighteen months to investigate the doses received by rats that burrowed into the hill. He found that the doses to the bronchial epithelium of the rat lungs were as high as 10,000 rem per year. This should be enough to cause lung cancer in the rats, and a number of them were trapped by Drew for pathological study, but no lung damage was found. This was not surprising, because a rat would be at a disadvantage with even slight lung disease, and thus would be more subject to predation. Hawks and snakes were likely to have found the diseased rats long before Drew did![13]

My Brazilian experiences revolved around persons with whom unusually close bonds developed. In addition to Roser and Cullen, there has been Eduardo Penna-Franca, who was one of Chagas's students when we started our collaboration. Penna-Franca eventually became director of the Institute of Biophysics, and the principal Brazilian investigator in our joint studies.

Roser and Cullen were classical Jesuit scientist-priests. Without the distractions of families, and with no materialistic interests, they were totally committed to the Church and science, and divided their long working days between their two spheres of devotion. They were good scientists, but I never made any progress in my many attempts to understand how they resolved the basic conflicts they must have faced. Although liberal on social issues, they were conservative in religious matters and, outwardly at least, experienced no difficulties in accepting the miracles that were part of their faith, while insisting on intellectual rigor when dealing with scientific matters. Roser was short in stature and a bundle of energy. He was an accomplished linguist who, when he became excited would speak rapidly and often switch unconsciously from one language to another. He had an excellent sense of humor, and had coined a number of delightful aphorisms. Among my favorites was, "We must do all things in moderation, especially those we do excessively!" The life of this accomplished scholar and priest was lost prematurely one morning when he was carried out to sea by an undertow while bathing at a beach in Rio de Janeiro.

Cullen succeeded Roser as chairman of the Physics Department and helped to build it into one of the best in Latin America. He loved field work, and we spent many enjoyable days in our explorations of the radioactive areas. When in Rio he would arise at four in the morning to offer mass at a convent some distance from PUC but said it was worth the effort because the nuns served ham and eggs that reminded him of his native Brooklyn. He possessed a fine tenor voice and had formed a choir that was much sought after for tours in Brazil and Europe. He was a gifted raconteur and displayed an excellent sense of humor. One evening we were at a bar in a small hotel in the state of Minas Gerais, where we had just explained to the bartender how to mix two gin martinis. When I asked him about the reaction of the townsfolk to a priest drinking in a public bar he noted that he was following the American custom of wearing trousers rather than the cassocks worn by Catholic priests in Brazil and most other countries in Latin America. "Why should I worry" he responded, "when they see my trousers they think I'm an Episcopalian"!

In the early days of our study, Cullen was ingenious in improvising the required scientific equipment. When a whole body counter was needed and he couldn't afford an expensive steel room, he contacted the Brazilian Sugar Institute and borrowed a hundred or more sacks of refined sugar which he stacked into a room. The sugar made a fine shield because it is free of radioactive impurities, but it was necessary to build it with a thickness of several feet because of its light weight. It had the additional disadvantage that the rats soon discovered its existence. At a more sophisticated level, he built a system for detecting the presence of thoron in human breath using a kitchen pot and scrap parts found around the laboratory. One of my satisfactions during my long association with PUC was watching the physics department at PUC develop from its original primitive condition to a well-equipped, modern laboratory. Unhappily, Cullen lost much of his effectiveness at an early age because of a series of illnesses and suffered a fatal stroke in 1985.

At about the time our studies were beginning, investigators in the U.K. discovered that Brazil nuts contained abnormally high concentrations of radium. This suggested that the locations where the nuts are grown in the Amazon Valley might also be a radioactive anomaly. Eduardo Penna-Franca arranged for the loan of a river boat from the Brazilian National Research Council so that we could travel up the Amazon and its tributaries to obtain samples for study. We were joined by Cullen and, with a crew of two and Eduardo's wife, Loi, acting as cook, we flew to Manaus, about a thousand miles inland, where we boarded our boat for the four-day trip. Much to our surprise, the area in which we found the nut trees had normal levels of radioactivity. When Penna-Franca returned to his laboratory he found that the high radium concentrations could be explained by the fact that the nuts had an affinity for barium, which is in the same chemical family as radium. The tree roots absorbed the radium because of the similar chemical properties of the two elements.

In 1980 I developed a new interest in the Morro do Ferro. I was then serving on the National Academy of Sciences Board on Radioactive Waste Management, where one of our concerns was the rates at which long-lived radioactive wastes migrate through rocks and soils and become a hazard to humans because of contamination of groundwater or agricultural products. Of particular concern to many people were the transuranic actinide elements such as plutonium, which has a half-life of about 25,000 years. It occurred to me that we could learn much about the long-term behavior of these elements in the environment by studying the rates at which certain elements migrate from the Morro do Ferro orebody, which has been in place for millions of years. The rationale for the study was that the chemistries of the rare earths and thorium are similar to those of plutonium and other transuranic elements. Since the orebody had been eroded close to the surface over geologic time, and was in a weathered state, it could be studied as an analogue for an ancient nuclear waste repository in which only the long-lived transuranic elements remained. The U.S. Department of Energy provided us with the initial research funds and the Brazilian Government cooperated in a major way by, among other things, establishing a field laboratory in Pocos de Caldas. Eduardo Penna-Franca and I served as co-principal investigators until my retirement, when Paul Linsalata succeeded me in the study.

Wayne Lei, one of our graduate students, went to Brazil to do his doctoral research on the Morro do Ferro project and to serve as our field supervisor. This required that he coordinate the logistic needs of the many Brazilian and U.S. specialists who became involved in the project. The research developed into a remarkably successful program in which we were able to demonstrate that the analogue elements were so well stabilized, that even under the adverse conditions that existed on the Morro do Ferro, they would remain in place, immobilized for millions of years by their attachment to the clay minerals in the soil. The project attracted international attention, and has been supported recently by several European countries.[14]

Lead 210 and Radon Exposure

Although the counter developed by Laurer was intended for plutonium, we realized that it would also be useful for other radionuclides that emit soft-gamma or X-rays. Among these was lead 210, a naturally occurring nuclide produced by the decay of radon. Lead 210 has a half-life of twenty-two years and decays by beta particle emission, but with an associated emission of a soft gamma ray. The beta particles can only penetrate a few millimeters of tissue, so that if deposited in the body they cannot be detected by external counting. However, we knew we could detect the soft gamma emission, and by that means we could measure the quantity of lead 210 deposited in the body.

This we recognized as important because it would give us a method of estimating the cumulative exposure of miners to the radon and its daughter products.[15] When radon or its short-lived daughter products are inhaled, they decay in a matter of minutes to the relatively long-lived lead 210 which then enters the blood. Lead 210, although radioactive, behaves like stable lead and is deposited in the skeleton, from which it is removed very slowly over a period of many years. The longer a miner is exposed to radon, or the higher the concentration of radon in the air he breathes, the greater is the amount of lead 210 accumulated in the skeleton. Thus, if the lead 210 content of the bones can be measured, it is possible to estimate the cumulative radon dose received by the miner. Such measurements, made in conjunction with epidemiological studies of lung cancer among miners, could help establish at what level of cumulative exposure the miner is at risk of developing lung cancer.

To test the concept, we arranged for four miners, all with known histories of heavy exposure, to be sent to our laboratory for measurement. The four men, two of whom were a father and son who both later developed lung cancer, came wide eyed to our laboratory. It was their first trip east, and they expected the laboratory to be located in the midst of what out-of-towners visualize as New York. Instead, they found themselves billeted at a comfortable conference center in the midst of a hemlock forest.

Lead 210 deposits in the skeleton, and since the skull is the largest mass of bone in the body and is covered by a minimum of intervening soft tissue, we placed the Laurer crystals at the head. After a thirty-minute count while the subject sat in a chair in our counting room, the spectrometer clearly demonstrated the presence of lead 210 in all four of the miners.

Aided by a grant from the U.S. Public Health Service and with the cooperation of the Department of Health in Colorado, we assembled equipment and staff for an expedition to the uranium mining areas of

Western United States in the summer of 1967. We made measurements of about a hundred men, both miners and control subjects with no history of mining, and found that lead 210 was readily demonstrated only in the miners. Some of the measurements were made in whole body counters located at the University of Utah and the Colorado State Health Department, but most were made in motel rooms in mining communities on the western slopes of the Rocky Mountains, where we successfully improvised methods of shielding our subjects from background radiation. The next step was to relate the size of the lead 210 "signal" obtained for each of the miners to the cumulative radon exposure of the miner. This required knowledge of the rates of absorption of radon and its daughter products into the blood, and the rates of deposition and removal of the lead 210 from the skeleton. With this information it would be possible to back calculate from a lead 210 measurement to dose to the lung. It was also necessary to calibrate the measuring system so that we could relate our instrument measurements to the amount of lead 210 in the skeleton. This required many months of laboratory work during which Norman Cohen, then a graduate student, studied the kinetics of absorbed lead in the baboon.

Pure chance continued to favor our research. At about the same time we were doing the lead 210 research, it was decided to move a major primate colony from Georgia to Sterling Forest. The primates were intended to serve the research needs of the medical centers in the New York area. Colonies of apes and monkeys were installed in a complex of specially designed trailers about one kilometer down the road from our laboratory. This made it convenient for animals to be transported to our laboratory where they could be examined in our specialized facilities. Cohen decided to use the baboon for his study, and was successful in measuring the distribution and elimination rate of lead 210 by serial measurements in the same animal, without the need for radiochemical measurements of tissues obtained either by surgery or during autopsy. This proved to be an excellent method of investigation, since measurements could be made in a species that had many metabolic characteristics that were similar to those of man, without sacrificing the animals or using invasive surgical methods.

Our studies demonstrated the feasibility of using lead 210 as a measure of cumulative radon exposure. I then proposed that the AEC conduct a survey of all uranium miners so that those with evidence of heavy exposure to radon could be identified, so as to reduce their risk of lung cancer.

They could be advised to leave underground mining and to quit smoking, since smoking has a synergistic relationship with radon exposure. They could also be provided with careful medical surveillance. All of these possibilities would have been of benefit to the miners. Unfortunately, in the mid-1960s the AEC was still of the opinion that uranium mine safety was not their responsibility. I experienced a feeling of *déjà vu* as I listened to arguments that it should be done by the states, the Department of Labor, or the Public Health Service. Twenty years before the AEC had taken the legalistic position that safety in the uranium mines was none of their business, and that decision cost five hundred lives.

Science and Bureaucracy in China

Among my many consultations in foreign countries, the most satisfying were those that resulted in collaborative research. The Brazilian relationships, which have covered a period of twenty-five years have been by far the most productive, but the three trips I made to China were the most interesting.

In 1980, shortly after relations with the People's Republic of China were resumed, I was invited by the Chinese Academy of Sciences to visit for one month. I made two additional trips in 1981 and 1984. Out of those visits came the opportunity to become involved with the studies then beginning in Yunnan Province, where more than 1300 miners had died from lung cancer resulting from their exposure to both radon and arsenic. My advice to the Chinese was based on straightforward principles of industrial hygiene and health physics, as a result of which five scholars of excellent quality were sent to our laboratory for varying periods of study and training. The visit also resulted in a formal collaborative agreement between the Institute of Environmental Medicine and the Chinese government. All of this made my trips to China worthwhile, but the most memorable aspects had to do with the opportunity to experience the cultural contrasts in a country that on the one hand was producing nuclear weapons and space satellites, while the civilian economy and most of the scientific infrastructure was comparable to those that existed in the West early in this century.

In 1980, I was one of the first Western environmental health specialists to visit China. When Irma (who accompanied me on the three trips) and I visited Geizu, where the tin mine was located, we learned that we were the first American visitors since World War II. I was the first foreigner to lecture at the Institute of Industrial Hygiene in Beijing, and the Institute of Radiation Protection in Shanxi Province. Very little English was spoken, and the quality of the interpretation was poor. A one-hour lecture easily stretched to three because of the need to slowly speak one sentence at a time and then wait for the Chinese interpretation. If there were members of the audience who also spoke some English, they would often interrupt the interpreter and argue about what I said. The slide projectors were primitive, and frequently caused delay.

At the time of my 1980 visit, the Institute of Radiation protection, which had a staff of 700, had no computers, no typewriters, and only three telephone extensions. I recall that one of the senior scientists took a piece of correspondence from me out of his pocket and asked how such beautiful letters were produced. However, this sort of contrast was not to last long. By the time of my visit in 1984, only four years later, most of the younger scientists spoke English proficiently, had computers at their disposal, and in many ways had closed the gap that existed at the time of my first visit.

Our progress in the Chinese investigations were repeatedly slowed by the extraordinary bureaucracy that existed. Routine decisions, such as minor changes in travel itineraries, took days to obtain. It took nearly a week for me to obtain a sample of dust from the rafters in the tin mine. The problem was illustrated when Dr. Sun Shi-quan, a pathologist from the Institute of Radiation Protection, visited our laboratory for one month, and I arranged a visit to major laboratories in the U.S. on his return trip. I suggested several laboratories and the scientists with whom he might wish to consult, all of whom were prominent and known to Dr. Sun from their publications. When we decided on a tentative itinerary, my secretary began to place telephone calls on the basis of which the visits were arranged in a few hours. Sun was amazed at the way this was done. In the U.S. there are both directories of scientists and telephone books that make it possible for anyone to call anyone else, facilitated by an efficient telephone system. My counterpart in China would not have had the authority to approve the visits in the first place, and since the individual visits would cut across administrative jurisdictions, he would have neither known how to make the contacts, nor have had the authority to do so. There has been a marked reduction in Chinese bureaucratic complexity since 1980, much of which is no doubt due to the influence of the thousands of young scholars who have been exposed to Western methods of scientific research, and have become less tolerant of red tape upon their return. (This was written before the violent events of 1989.)

One of the problems at the tin mine was that there was great uncertainty about the cumulative exposures of the miners to radon. With funds from the U.S. National Institute, Laurer constructed a lead 210 counter in the mining community, where measurements of the skeletal burdens of the miners are being made.

HUDSON RIVER STUDIES

The move to the forest opened a number of opportunities for ecological research that would not have been practical in Manhattan. In particular, I was impressed by the advantages of our proximity to the fresh water reaches of the Hudson River.

The Hudson is one of the major rivers in the eastern United States. It rises from tiny Lake Tear of the Clouds in the Adirondack Mountains and flows almost inconspicuously to a dam at Troy, still 150 miles from the sea. At the Troy dam the river bottom is four feet below sea level, so that the river is subject to tidal influences below the dam. But the water remains fresh until the vicinity of Poughkeepsie, about seventy miles below the dam and eighty miles from the sea. For about fifty miles between West Point and the George Washington Bridge connecting New Jersey with Manhattan, the river flows past the Hudson Highlands, the grandeur of which attracted the Hudson River school of painters in the nineteenth century. Earlier, the beauty of the highlands attracted the attention of Dutch and English colonials, some of whom received large grants of land that were managed with wisdom and foresight and eventually donated to the states of New York and New Jersey to be preserved as parks that have become models of land and forest preservation. Roosevelt, Rockefeller, and Harriman are some of the names associated with large tracts that have been preserved in this way. The Hudson channel can be traced far out to sea.

Like other great rivers of the world, the Hudson serves many purposes. It is a major artery for transportation, provides water for domestic and industrial use, serves as a drain for wastewater, and is an unsurpassed recreational resource in an area of great natural beauty and important historical associations. Yet despite the importance of the river, in the early sixties we knew hardly anything about the ways in which it functioned as an ecological system.

In 1962 the first nuclear power plant built by Consolidated Edison at its Indian Point station on the east bank of the Hudson River went into operation. At about the same time, I became interested in studying the fate of the vast quantities of nuclear-weapons fallout that were deposited on the 13,000 square mile Hudson River drainage basin. Some of the radionuclides, such as strontium 90 and cesium 137 had half-lives of about thirty years, and would not soon decay. Would they remain in place in the soils and sediments, or would they mobilize and enter the food chains, eventually to be absorbed into the human body? To what extent would some of the radionuclides dissolve and enter drinking water? These were interesting questions deserving of scientific study, and also had important public health implications. Accordingly, I discussed the need for a radioecological study of the Hudson River with both the New York State Health Department and the U.S. Public Health Service. They agreed to support a pilot program.

The team we assembled to undertake the Hudson River studies was strengthened by the inclusion of faculty and graduate students from the biology and geology departments of the NYU Washington Square College. One of the biologists, Alfred Perlmutter, had participated in a study of Hudson River ecology thirty years earlier, and I was surprised to learn there had been no studies since. The New York State Department of Conservation was so eager for us to begin our studies that they loaned us the services of a boat and crew. This proved invaluable during our initial field trips.

Shortly after our studies began I attended a conference at Hanford, Washington, where I met Gwyneth Parry Howells, an English ichthyologist and well-known specialist on the physiology of trace substances in aquatic species. When I told her about the studies we were planning, she expressed an interest in joining our group on her sabbatical. This I quickly arranged. She became an important member of our faculty and remained with us for five years before returning to the marine biology laboratories of the Central Electricity Generating Board in Great Britain. In addition to directing the Hudson River studies during its early years, Howells collaborated with a number of graduate students and faculty in studies of the physiological kinetics of trace substances.

I had lived near the shores of the Hudson for most of my life and, like most New Yorkers, had developed an affection for it. As a child my parents had frequently taken me for outings on the excursion steamboats that plied the river to Bear Mountain and West Point. Riverside Drive and the narrow park it bordered, stretching for five miles along the west side of Manhattan, was convenient for biking, roller skating, and walking. We pushed our three boys for miles along that walk in baby carriages. For a few cents we could walk across the George Washington Bridge and, once on the New Jersey side of the river, hike north through the Highlands for days. The prospect that I would be organizing studies of the Hudson River ecology was personally very exciting.

When our studies began we were astonished that so little was known about the ecology of the river. There had been no systematic attempt even to catalogue the flora and fauna of the river since the 1930s, when a modest inventory of the indigenous fish and rooted plants was compiled by scientists supported by the Works Progress Administration. That research was performed during a single summer, and was of uncertain relevance to the conditions that existed about thirty years later.

Our first task was thus to learn about the abundance, distribution, and variety of the animals and plants in the river, from microscopic forms to fish. For 25 years, from 1962 to the present, the staff and students of the laboratory have been towing seines up and down the river, gathering samples of the hundreds of different species of lower life forms as well as fish for study in the laboratory.

We had been led to believe that the Hudson was a "dead" river because it was used as an open sewer for untreated municipal and industrial wastes. In fact, the river supported an active striped bass and shad fishery. We did find that there were fewer fish species in the river than had been reported three decades earlier, thirty-five compared to seventy. We found no sturgeon, an important food fish in earlier years that had sometimes grown to seven feet and attracted the attention of river travellers because of the habit of occasionally leaping from the water. However, apart from the diversity of fish species, the health of a river must be judged also by such matters as the oxygen content of the water, the presence of nutrients as well as toxic chemicals, and above all by the number and diversity of the many planktonic species which comprise the ecological web on which the fish, who are at the top of the aquatic food chain, depend for their existence.

When we decided to begin the Hudson River studies in 1963 we did not know that a book published that year would have profound effects on popular attitudes towards the quality of the natural environment. The book, *Silent Spring* by Rachel Carson, described in prose of great beauty the delicate ecological balance that exists between all forms of life and warned about the dangers of allowing toxic organic chemicals such as the synthetic pesticides to contaminate the environment. The modern environmental movement began when it did because of the impact of that book on the thinking of scientists, legislators, industrial executives, and the general public. The word "ecology" burst into the popular vocabulary with great force and made the public aware of the reciprocal relationships that exist between human society and the environment on which it depends for existence.

Our small group thus found itself beginning a scientific investigation of Hudson River ecology just when popular interest in the subject was developing for the first time. Our plan was to catalogue the flora and fauna of the river and investigate the mechanisms by which various radioactive substances, either of natural origin or present in fallout, pass through the food web and eventually reach organisms consumed by man. However, our efforts were soon diverted to other problems of ecological impact, of which several important ones directly involved power generation. These were, first of all, the effects of drawing huge quantities of river water into the proposed pumped storage facility at Storm King Mountain; secondly, the effects of warm water discharged from power plant condensers into the river and finally the effects of the severe pressure and temperature regimes on organisms entrained in river water circulating through power plant condenser cooling systems.

The Storm King controversy was one of the first in which a major construction project was blocked by an aroused public. The proposed development was a pumped-storage power plant based on the principal that the economics of electrical generation is improved if power stations can operate around the clock rather than only during the hours of peak demand. Since demand is lower during the night hours, the economic performance of the plant can be improved by using the excess generating capacity during the slack hours to pump river water into an elevated reservoir from which it can be released through electric turbines when needed. Con Ed proposed that the plant in question be located at Storm King Mountain just below West Point on the west shore of the Hudson River. The top of the mountain was to be carved into a storage reservoir. Power brought to the base of the mountain by transmission lines would operate pumps to raise water into the storage reservoir, from which the water would return to the river as needed, after passing through turbogenerators that would return the "borrowed" energy to the power grid.

In retrospect it is likely that the original intervention was motivated less by ecological considerations than by the understandable wishes of well-to-do landowners to preserve the appearance of the landscape. However, the main theme developed by those who sought to block the development was that the eggs of larvae of striped bass would be destroyed in passing through the system. This controversy was one that our laboratory succeeded in avoiding, although Al Perlmutter of the Biology Department became involved in his individual capacity, because of his unique knowledge of the life history of the Hudson River striped bass. Con Ed finally gave up the Storm King project after many years of expensive litigation.

The ecological effects of the warm water discharged from power plants was another subject that attracted considerable attention during the mid-1960s.[16] Power plants that are near sufficiently large bodies of water have traditionally pumped the water through condensers that remove waste heat from the turbines. The water used for condenser cooling is then returned to the source from which it was drawn, but its temperature is elevated. This "thermal pollution" became a major source of concern because little or nothing was known about the ecological effects of the warm water. The warm water attracted sport fish such as striped bass, so that many power plants found that it was good public relations to build fishing piers near the discharge point for the warm condenser coolant water. But nothing was known about the effects of the abnormally high temperatures on the growth and development of the lower forms of life on which the ecological health of the aquatic environment depend.

A related problem was the effects of the severe stresses placed on aquatic organisms when they are entrained in water that enters the cooling system. Although the water enters through intake screens designed to keep flotsam from entering the condensers, the velocity through the screens was so high that mature fish were sometimes impaled against them in large numbers. Before the emergence of an environmental consciousness, this was often a source of amusement among power plant employees and local residents who would come to the condenser cooling system intakes to fill baskets with fine specimens of game fish picked from the intake screens. The mature fish so entrapped represented an insignificant form of predation, but what about the millions of eggs, larvae, and planktonic organisms. The condenser cooling systems recirculated large volumes of water so frequently that it was possible to calculate the high probability that an organism would be entrapped in this way. The ecology of the river could be seriously disrupted if those organisms were killed in transit through the cooling system by the high temperatures and pressures.

Our laboratory soon received a research grant from Con Ed to study the environmental impact of condenser cooling systems. The staff's basic technique originally consisted of placing seines at the intakes and discharges of the cooling system. The number and kinds of organisms found at the intakes were then compared to the numbers of viable organisms found at the discharge, with the difference attributed to the effects of passage through the condensers. This required difficult and dangerous sampling procedures which I observed many times, often in the middle of the night when certain organisms were more abundant than in the more convenient daylight hours. One night it occurred to me that the entire process could be readily simulated under laboratory conditions by constructing a simplified condenser in which one tube, constructed to full scale, could be heated and pressurized as desired. The power plant provided the worst possible conditions for study because the desired variables could not be controlled. Only with a properly instrumented simulator could the effects of entrainment be studied properly.

The next morning I discussed the idea with engineers of the New York State Energy Research and Development Authority, as a result of which we received a grant to build the simulator and undertake the required studies. This was only one of many examples in which we succeeded in orchestrating private and governmental sources of support for the multidisciplinary kinds of research we were doing.

The simulator was built and allowed one of our graduate students to conduct his doctoral research, which provided much needed information on the ecological effects of condenser cooling systems. The information obtained made it possible to lessen the ecological impact by properly controlling the biocides which were injected into the cooling system to prevent fouling of the condenser by various species of aquatic life. It was also learned that the organisms could tolerate changes in pressure and temperature provided that cavitation (creation of bubbles) was avoided.

The support our laboratory received from Con Ed presented me with a dilemma because I had been a consultant on radiological safety matters to the company for several years before they began to support our research. For the consultations I received fees which, according to general university practice, I was permitted to retain. Serving as a consultant to a private company while accepting research grants from it could be perceived as a conflict of interest. Twenty-five years ago, however, whether a conflict actually existed was judged on a case-by-case basis. Today, industrial companies are frequently denied the opportunity to obtain expert advice from academic scientists because they wish to avoid even the appearance of a conflict of interest even where a conflict does not exist. I would not enter into such an arrangement now because of the much greater sensitivity to the appearance of conflicts of interest. As events developed, I was the one who broached the subject of concluding Con Ed's support for the laboratory because the research was becoming burdensome. The company had become involved in litigation that required collecting information on a scale and of a type that did not seem an appropriate task for a university research laboratory. After an interval that allowed Con Ed to find another laboratory, the connection came to an end.

Research on the original goal of understanding the fate of radionuclides deposited on the Hudson River watershed has continued for more than twenty-five years. It has been a small but productive program which has tested models that predict the dose to people from discharges of reactor emission into the river. Paul Linsalata, who was one of the few graduate students appointed to the faculty after receiving a Ph.D., found that the models published by the regulatory agencies tended to overestimate the dose received. He and the students who preceded him also learned that most chemical and radioactive pollutants do not remain suspended or dissolved long in the river water, but move to the sediments from which they enter the aquatic food chains.[17]

Rachel Carson's book called attention to the persistence of certain synthetic organic chemicals in the environment. Much of the Hudson River research concerned the measurement of the sources and effects of polychlorinated biphenyls (PCBs) in the river. PCBs are an important group of chemicals used in industry for many purposes. The extreme toxicity of one member of this family of chemicals, Halowax, when used to insulate electrical cables was discussed earlier. The PCBs have been used to provide electrical insulation in electrical equipment of many kinds as well as dielectrics in condensers. The dangers of using PCBs came to worldwide attention as a result of a 1968 accident in Japan, in which rice oil became contaminated by a leak in a heating system. Many of those who consumed the rice oil became ill and many died, mainly as the result of damage to the liver.

Just prior to this episode, our laboratory was one of many that had begun to make measurements of the DDT family of insecticides which were persistent environmental pollutants that became progressively more concentrated as they passed from water or soil to higher levels in the ecosystem. At each step in the food chain, they can be concentrated ten to one hundred times the concentration in the previous step. Both PCBs and DDTs tend to accumulate in the fatty tissues of plants and animals. At the top of the food chains, in fish-eating birds for example, the amount of contaminant per gram of tissue can reach ten million times the concentration in the contaminated water. DDT proved to be particularly toxic to birds who were exposed by eating fish and insects.

Although the environment was known to be contaminated by a number of chlorinated organic chemicals, it was difficult to demonstrate their presence quantitatively until the 1960s when advanced systems of measurement were developed. In particular, the gas chromatograph, especially when used in conjunction with mass spectroscopy, made it possible to separate and quantify the individual members of these complicated families of chemicals. Much to our surprise, when the instruments used in our laboratory reached a sufficient level of sophistication, we learned that what we had thought were DDTs in the samples of sediments and biota collected from the river were in fact PCBs. Following this finding, Joseph O'Connor, the head of our river program, began an important series of studies of how the PCBs behave in a river system and the rates at which they accumulate and move through the ecosystem.

Contamination of the Hudson by PCBs eventually developed into a cause célèbre. A major source of PCB contamination was the release of about thirty pounds per day from two General Electric plants that manufactured electrical condensers in towns on the upper reaches of the river. Over a period of many years, the PCBs accumulated in the sediments upstream from a dam, whose removal caused the contaminated sediments to pass to the lower river and its biota. When the extent of contamination of fish became apparent, the New York State Health Department forbade commercial fishing of striped bass and shad, both of which had long been taken by commercial fishermen. It was ironic that at a time when the ecological health of the river was improving as a result of strict prohibitions on the release of domestic and industrial pollutants, it became necessary to discontinue commercial fishing because of the presence of a persistent contaminant that was likely to be present for many years.

Another significant source of contamination was a small cove on the east side of the river opposite West Point, where the sediments were contaminated by cadmium released from a factory that had produced batteries during World War II. Theodore Kneip and his graduate students found that the metal was being selectively sequestered in the hepatopancreas of the blue crab, a much sought after delicacy. The hepatopancreas is the organ known as the "mustard" by the blue crab fanciers, but because of its tendency to concentrate cadmium, the Health Department issued a warning to the blue crab catchers that the "mustard" should not be consumed.

At this point it is relevant to note that my understanding of the importance of the Hudson in the long-range needs not only of New York City, but the entire river basin, was substantially improved by my service as Environmental Protection Administrator of New York City during the period of 1968–1970. While most of the events of those two years are in the next chapter, the knowledge I gained about the river is particularly relevant here. Questions with enormous financial implications arose that could not be answered rationally because not enough was known about the river as an ecological system. Among the striking examples during my term:

1. A 200 million-gallon wastewater treatment plant was scheduled for construction but the design criteria were uncertain because not enough was known about the ecological characteristics of the river. This involved a 500 million-dollar expenditure.
2. Did the Con Ed plants at Indian Point require cooling towers, or was the cooling capacity of the river water adequate to cool the condensers? This was a 400 million-dollar decision that was also difficult to resolve because of insufficient information.
3. The metropolitan area will some day be required to consume Hudson River water for domestic purposes. How should the river basin be managed during the decades ahead to insure that the water will be of adequate quality when needed? This question will involve not only expenditures of billions of dollars, but will also affect the health of tens of millions of people.

It became apparent to me that the river as an ecosystem must be protected for all time. To assure its ecologic health, yet satisfy the needs for economic development of the Hudson River Valley would require scientific information that could not possibly be obtained without great amounts of money for research and monitoring. I discussed the matter with Commissioner Henry L. Diamond of the New York state Department of Environmental Conservation, as a result of which I was asked to chair a committee to study the matter further. Other committee members were Dwight Metzler, Deputy Commissioner of Environmental Conservation, Martin Lang, Commissioner for Water Supply of New York City, and Charles E. Rider, Vice President of the Central Hudson Gas and Electric Company.

In October 1972, we submitted our report to Diamond in which we confirmed that the required studies would involve research support on a level far beyond what is available from conventional sources. We estimated that the research needs could easily justify an investment of $20 million per year (in 1972 dollars) for many decades. Our report concluded that when the research needs are so intimately connected to the future well being of so wealthy and thriving a region as the Hudson Basin, the required funds should be provided as a proper charge against future demographic and economic development. Based on the projected cost of new construction in the region, we recommended that the funds be obtained by a tax of 0.7 percent on construction. We also suggested ways to administer the funds to assure that the research grants would not be subject to political influence and that the results of the research would be of high quality. Diamond and his associates were impressed with the report and the concepts it introduced. It was presented to Governor Nelson Rockefeller who gave the recommendation his support, and requested that drafts of the required legislation be prepared.

I was of course elated with the prospect that we might have developed a concept by which the future health of the river could be better planned. Unfortunately, our well-developed plans were thwarted by tragic developments at a higher level of government. In Washington, President Nixon resigned from office under threat of impeachment and was succeeded by Gerald Ford who summoned Nelson Rockefeller to Washington as his vice-president. Diamond, who was a close advisor and friend of the governor, resigned from New York state government and went to Washington. From time to time during the past fifteen years, I have attempted to revive interest in the plan but I have not succeeded. This is unfortunate because the concept is still viable.

Despite this disappointment, some good did come out of our study: the New York state electric utilities have established the New York State Electric Research and Development Authority (NYERDA) which conducts a program of research needed by utilities. Some of that money is invested in Hudson River research, but not enough.

Shortly after I retired from New York University in 1985, I was pleasantly surprised to receive the first Distinguished Service Award given by the Hudson River Environmental Society in recognition of my contribution to the preservation and management of the river. This pleased me very much.

Studies of Urban Air Pollution

When the first commercial nuclear power reactors began operation in the early 1960s, the public was concerned about the risk presented by radioactive gases the reactors would emit to the atmosphere. The amounts were insignificant in terms of the dose delivered to the public, but it was repugnant to many to think that any radioactive materials were being released into their neighborhoods. One day it occurred to me to make an estimate of the amount of radioactive dust emitted to the atmosphere from coal-burning plants. I first obtained from the literature the content in coal of radioactive elements like uranium, thorium, and potassium. I was surprised to find that the dose received by people living in the vicinity of coal-burning power plants might be higher than that received near pressurized water reactors. The dose from the coal-burning plants was of course very small, but the dose to people living near the reactors was even less. This judgement was first made from a few notes scratched on a piece of paper, but seemed to be sufficiently important to confirm by actual measurements. I called my contacts at Con Ed, and without telling them what I had in mind, I asked them for samples of the "fly ash" routinely discharged from their coal-burning plants. I then gave the samples to one of our advanced graduate students, Henry Petrow, and requested that he analyze them for certain of the more important radionuclides that occur naturally in fossil fuels. His results confirmed the values I had found in the literature, whereupon I assembled the information into a short paper which we submitted to the journal *Science*. The paper was quickly accepted for publication and has helped to place low-level radiation exposure into perspective.[18]

In 1968 I had reason to become even more interested in air pollution in a more general way while I was Environmental Protection Administrator of New York City. The public was clamoring, rightly so, for clean air, and one of my major responsibilities was to mount a major program of air pollution control. Although I was on leave from NYU, I was successful in gaining research funds for the laboratory to develop methods to apportion the relative contributions of various sources of emissions to the pollution in the city's air. The details of the air pollution problem and the results of the research are in a subsequent chapter, but it should be noted that scientific research is important for public policy, even on a local scale.

Retirement from NYU

By 1984, except for my two-year leave of absence to assist the City of New York, I had devoted twenty-five years to the Institute of Environmental Medicine of NYU. I had reached the age of sixty-nine and, in conformity with university practice, I would be required to relinquish my responsibilities as director of the Laboratory for Environmental Studies in another year.

Irma and I discussed the options available and decided that the time had come to relocate, preferably to a milder climate. Although we very much enjoyed life in Sterling Forest, the winters were severe and seemed increasingly so with each passing year. We selected Chapel Hill, North Carolina for our new home. It is a college community, the home of the University of North Carolina, and is only a short distance from Duke University in Durham and North Carolina State University in Raleigh. The three universities form a triangle of rolling country, part of which has been developed as a research center where several government and private laboratories are located. Chapel Hill would have been an ideal retirement community in its own right but proved especially attractive when, at the initiative of two colleagues, Arthur Stern and Leonard Goldwater, I was offered courtesy appointments on the faculties of both Duke and the University of North Carolina.

In September 1985 we left our home on Sterling Lake, where we had lived pleasantly for nearly a quarter of a century, and moved into a delightful setting on the edge of the Arboretum of the University of North Carolina, about two miles from the campus and the center of town. Yet another pleasant experience remained for me at NYU. At the suggestion of a number of my former graduate students, most notably Peter Freudenthal, a graduate fellowship program was established in my name. The proposal drew widespread support, resulting in a fund whose income supports worthy doctoral students at the institute. During the summer of 1989, I returned to Sterling Forest to meet the first two students to be awarded Eisenbud Fellowships.

EIGHT

Environmental Protection Administrator, City of New York

In January 1968, I took leave from NYU to accept a two-year appointment from Mayor John V. Lindsay to form an Environmental Protection Administration for the city of New York. This was a novel idea conceived in Lindsay's first administration (1965–69). To his credit, Lindsay was among the first to recognize the need to unify the administration of the many branches of government responsible for water supply, wastewater treatment, air pollution control, noise control, and general sanitation, including refuse collection and street cleaning in an Environmental Protection Administration (EPA). New York City's EPA preceded that of the federal government by about three years.

These several functions were traditionally spread over many departments of the city government, often with conflicting objectives. For example, the Department of Sanitation decided in the early 1950s that new apartment houses should be required to incinerate their garbage so as to reduce the volume of refuse. This may have seemed sensible to the commissioner of sanitation, but it created a major new source of air pollution. By 1968 there were thousands of badly operated refuse incinerators, each spewing smoke and gases into the environment. In another example, the department purchased garbage collection trucks that were well suited to its purposes, but were sources of a shattering racket. The department's job was to collect and dispose of garbage, not to worry about noise. These were the kinds of problems that Lindsay's advisors correctly perceived could be minimized by combining all city departments concerned with air, water, solid waste, and noise pollution into a single agency called the Environmental Protection Administration. The federal government and many states subsequently followed the same reasoning.

It was planned that the first EPA administrator would be James L. Marcus, one of Lindsay's political associates who was serving as commissioner of water supply, gas, and electricity. Marcus had started as an unpaid assistant to Lindsay soon after his election in 1966. Apparently little was known about him except that he had married into a socially prominent family.

A question that immediately arises is why Marcus had been given that important post, which carried major responsibilities that were hardly less than the position of EPA administrator. The biographical summary he had submitted to city hall inflated his educational accomplishments, but he did not claim to have experience that qualified him to be in charge of one of the world's largest water supply systems. The answer is simply that major positions in government at all levels are frequently filled by political appointees who have neither the technical background or experience demanded by the position.

In December 1967, Marcus was arrested for soliciting and accepting a bribe in connection with a contract to refurbish one of the city reservoirs, a crime for which he was later given a jail sentence. Those were the circumstances behind the call I received in early January 1968 from the mayor's brother, George Lindsay, who was assisting him in recruitment. The arrest of Marcus evidently had a profound effect on Lindsay, who needed candidates for both the EPA position and the position of administrator of the Health Resources Administration. He decided not to draw on his pool of political supporters, but to find professionally qualified candidates.

The title "administrator" was newly created. The bureaucracy of New York City had traditionally been organized into functional departments, each headed by a commissioner, most of whom were political appointees. There were a few exceptions, such as the Departments of Health and Police, in which it was recognized that the commissioner should have special qualifications. The city bureaucracy was so large that more than eighty departments, boards, and commissions reported directly to the mayor. Because this was far too large a number, Lindsay combined departments with related functions into "super agencies" called "administrations." Each of the nine super agencies was headed by an administrator.

The EPA combined the Department of Sanitation, the Department of Air Resources, the Bureau of Water Supply (formerly within the Department of Water Supply, Gas, and Electricity), and the Bureau of Water Pollution Control (formerly located within the Department of Public Works). It was a huge organization, with about 20,000 employees, an operating budget of $275 million, and, as it developed, a construction program of about $2 billion over a five-year period. In addition, a Bureau of Noise Control was created and located within the Department of Air Resources.

The reorganization plan was logical but encountered several serious problems from the outset. One was that some of the commissioners resented their loss of direct access to the mayor. This of course was the very problem he was trying to correct: too many people were coming to him for decisions that could be made by the administrator, who served as a sort of assistant mayor. Second, the legislative branch of city government, with control over the purse strings, couldn't understand why more high priced bureaucrats were needed. Many members of the city council looked upon the newly created positions as simply additional patronage for the mayor.

Strangely, they were partly right. Lindsay knew that a reorganization was needed and also that he needed to attract top-level professionals, but he was inconsistent in that he often filled posts with political appointees, as in the case of James Marcus. He used good judgement in some of the other administrations, such as the Human Resources Administration for which he selected as administrator Mitchel Ginzberg, the distinguished dean of the Columbia University School of Social Studies. Lindsay was also considering Bernard Bucove, commissioner of health for the state of Washington, for the position of health services administrator at the time of his discussions with me. Bucove had doubts about accepting the position because of the Marcus matter. He was interested in whether I would accept an appointment because he thought it would indicate the kind of people Lindsay was seeking for his inner cabinet. Lindsay was aware that Bucove was delaying his decision until he learned what I was doing, so he was all the more anxious to have me come aboard. When I announced my acceptance, Bucove did too. The two of us enjoyed a rewarding and productive professional relationship during the two years I was in office.

I saw the Lindsay offer as a challenging opportunity, but had no desire to remain in public service for more than two years. The mayor recognized that this was sufficient time to allow me to organize the EPA, after which I could return to Sterling Forest. On that basis I requested leave from NYU for two years. Norton Nelson and President Hester arranged for my leave on short notice.

Although I accepted the appointment in late January 1968, it was agreed that I would not assume office until March 15. This would permit me to wind up my affairs at NYU, take a brief holiday, and still leave me ample time to orient myself to my future responsibilities. For more than a month I was able to meet quietly with civic leaders, the mayor and his staff, and the appointed officials for discussions and briefings about the problems I would face. This proved to be a wise way to get started: I could not possibly have found the time for such in-depth discussions after assuming office.

The function of environmental protection in government was a new concept and, to my knowledge, I was the first person anywhere to hold the title of "environmental protection administrator" when I took my oath of office on March 15, 1968. Mayor Lindsay officiated in the well-attended ceremony at which many members of my family as well as friends and professional associates were present.

During the next two years I spent long days and many nights taking care of my responsibilities, but I never lost contact with the laboratory, where I spent my weekends whenever possible. The NYU Washington Square Campus is located on the northern edge of Greenwich Village, a thirty-minute walk from city hall. The university made available a small but pleasant apartment on the thirtieth floor of an apartment house they owned on Bleecker Street, and there Irma and I spent all of our weekdays and many weekends during the next two years.

Every Tuesday morning at eight o'clock I attended a meeting of the "super cabinet" in the basement of Gracie Mansion, the mayor's residence. This was usually a productive meeting of about twenty of the top officials and mayoral assistants. On Friday mornings I attended a meeting of about seventy-five commissioners and lesser mayoral aides in city hall. This was the "cabinet," in contrast to the smaller super cabinet. It was an excellent demonstration of the need for the reorganization. On Tuesday morning we heard progress reports on the state of the city, and enjoyed a productive dialogue with the mayor. On Friday at the larger meeting we could do no more than listen to one or two prepared presentations.

SOLID WASTE MANAGEMENT

New York City, like all municipalities, even more now than in 1968, has been facing increasingly serious problems caused by the burgeoning volumes of solid waste. The city has six thousand miles of streets. The resident population of more than seven million people is joined each workday by more than two million commuters. Each of the millions of people—residents and commuters—generates solid waste at a rate of about six pounds per day. This results in three basic problems: how to keep the streets free of litter, how to collect the refuse from homes and commercial establishments, and where to put it. Refuse collection by the New York City Department of Sanitation is limited to city-owned buildings and private homes and apartment houses. The commercial establishments are serviced by private carters.

Solid waste management in general was the responsibility of the Department of Sanitation, which was organized along military lines and had a long history of chronic problems that seemed to defy solution. Not the least of these was that the department had repeatedly been involved in scandals. It was also the only department in city government in which the increases in wages over the years had not been associated with corresponding increases in productivity. The only basic change in the technology of garbage collection in this century had been that the horse was replaced by the internal combustion engine. Finally, there had been a tradition of independence, defiance, and labor militancy. The 10,500 members of the Uniformed Sanitationmen's Association were led by John DeLury, a man of varied but fully controlled temperament who could change in an instant from a music-loving, courteous, charming, and seemingly considerate colleague into a raging, profane bully.

When I accepted the position of EPA administrator in January 1968, there were ominous indications of stormy weather ahead. The union's contract with the city had expired on June 30, 1967, and a strike was inevitable. The only question was whether it would be over by the time I took office on March 15.

Toward the end of January, Irma and I joined my sister and brother Elsa and Leon and their spouses Sid and Naida on a ten-day vacation in the Virgin Islands. On February 2, shortly after our arrival, the radio carried the news that the sanitation workers were on strike. Five days later DeLury was jailed for defying a court order to resume collections. His power, even while in jail, was illustrated when, out of compassion, he ordered five hundred of his men to resume refuse collections at hospitals and schools, which they did without pay.

I called the mayor's office from Virgin Gorda to ask if I should return to New York because of the emergency. Not that there was anything I could possibly do to help, but I thought that just being on hand in the background might give me some insights that could prove helpful in the future. I was delighted to be advised that I would only be in the way and should enjoy what would likely be my last opportunity for relaxation for quite a while.

The strike lasted for nine days, during which time more than 100,000 tons of garbage accumulated on the streets. A health emergency was declared, which the mayor attempted to remedy by having 3,000 civil service workers assigned to garbage removal duty. When they refused, Lindsay suggested that Governor Rockefeller call out the National Guard, but by that time it was clear that a long smoldering political rivalry between the two officials was complicating the matter. Rockefeller wisely refused to call out the guard, which probably would have resulted in a general strike. However, Rockefeller did enter the negotiations, and the men returned to work at his request. In the end, the union was given most of its demands, which Lindsay and the newspapers thought were exorbitant. The bitterness caused by the strike affected the future relationship of the governor and the mayor no less than that of the mayor and the union.

The strike occurred while the Department of Sanitation was without a commissioner, which was not an unusual condition during the Lindsay regime. The post had been filled by a succession of political appointees, the last commissioner having been Samuel Kearing, who had resigned angrily several weeks earlier. Maurice Feldman, whom I later appointed commissioner of water resources, was acting commissioner during the strike, doing an impressive job under very difficult conditions. When I returned to New York after the strike had been settled, I asked Feldman if he would accept a permanent appointment as commissioner of sanitation, but he declined. I discussed the kind of individual who was needed with him and others and concluded that someone with a military background and command experience in the management of large contingents of noncombatant personnel should be sought. I turned to the U.S. Navy, and interviewed several recently retired officers from the SeaBees, the navy's construction battalions. The SeaBees were known for their ésprit de corps, which meant that they were well managed despite the difficult conditions under which they frequently were required to work. Captain

Griswold Moeller, who had recently retired from active duty, came to my attention. I was delighted when he agreed to take on the challenges of heading the department. Moeller served with distinction throughout my term, but left office shortly after I did.

One of the problems I faced was that John DeLury did not accept the fact that his men were administered by a commissioner who did not report directly to the mayor. DeLury was obsessed with the importance of parity among the three uniformed departments. The commissioners of the other two uniformed forces, the firemen and the police, reported directly to the mayor, and he wanted the commissioner of sanitation to report directly as well. Lindsay was insistent that the Department of Sanitation remain in the EPA, although one concession was made: Moeller was included in the Tuesday morning meetings of the super cabinet, as had been the commissioners of fire and police. To DeLury's credit, he did raise the public image of his men, despite the fact that their educational requirements were lower than those of the two other uniformed forces.

Moeller was determined to improve the efficiency of the department. Metal cans had been an obstacle to improving collection efficiency for more than a century. Each sanitation truck required a crew of three, one of whom stayed in the cab to inch the truck from door to door along its route. The other two men followed on foot, collected the garbage cans one at a time, shook their contents into the truck hopper, and then noisily returned the cans to their original position. Plastic bags, of the type that are very common nowadays, had recently been developed and it was suggested that they might replace the cans. This would eliminate a source of noise as well as greatly increasing the productivity of the sanitation crews. One man could easily pick up two bags and quickly drop them into the hopper. The bags would also ease the work of the men because they were lighter than the cans and there would no longer be a need for strenuously shaking the cans to dislodge their contents.

It sounded like an excellent suggestion, but there were objections. The Health Department was concerned that rats, cats, and dogs would chew their way into the bags. The Fire Department worried that the bags of garbage would be ignited by cigarettes tossed into them by careless smokers. After much discussion it was decided that the idea was worth a trial. The Sanitation Foundation of the University of Michigan was retained to conduct a testing program, which the bags passed successfully. Plastic bags are now used for refuse disposal worldwide. They have not only improved the efficiency of refuse collection but have also eliminated a source of racket.

The bags were popular with both the sanitation men and the public, and they quickly displaced the cans. I had hoped that the union would cooperate in this opportunity to improve productivity, but they did not readily agree. Moeller wanted to change the collection procedure, because the plastic bags made it possible for one man to do the work of two, with the other man following with a broom to sweep litter from the curb. During our tenure we did not succeed in obtaining the cooperation of DeLury, so no changes were made in the way the three men were used.

Like most cities throughout the world, New York has been filling its lowlands with refuse for centuries. About 10 percent of the city, including some of its most desirable areas, has been created with garbage dumped into marshes and swamps or along the shorelines of the five boroughs. In 1968, each day the department collected about 25,000 tons of garbage which were disposed of either in landfills or by incineration. A problem we faced (and the city still faces), was that the landfills were filling up, without any clear alternatives for disposal. To reduce the volume of waste, the city had built eleven large incinerators but these were of obsolete design and major sources of air pollution. One of the first actions I took, on the recommendation of Commissioner of Air Resources Heller, was to shut down three of the incinerators owned by the city.

This was done because the incinerators were in no condition to be upgraded and also were in parts of the city where they were major nuisances. By this action, the pressure on the remaining landfill sites was increased, despite the fact that they had only limited capacity. To extend their life, I recommended that we increase their height and create hilly topographies wherever possible, rather than flat surfaces. This we decided to do, leading one of the morning newspapers to suggest sarcastically that my plan would make the landfills suitable for year-round skiing down the slippery garbage slopes!

The problem of refuse disposal defied solution when I was confronted with it in 1968, and the situation has not improved since then. We badly needed additional incineration capacity, but not everyone was certain that the resulting air pollution could be controlled. We did decrease the air pollution produced by the incinerators we kept in operation by fitting the stacks with scrubbers and electrostatic precipitators, but that was only a partial solution to the problem. Sam Kearing had conceived of constructing a large incineration unit next to the Brooklyn Navy Yard, across the East River from lower Manhattan. It would produce steam which Con Ed would purchase and pipe under the river to Manhattan to heat buildings. This was a thoroughly sound proposal, which made sense both economically and ecologically, since the waste would produce usable energy. We also began to plan for a second huge incinerator in the South Bronx.

There was every reason to believe that the new incinerators would not pollute the atmosphere, but opposition arose from local citizens who did not share our convictions. When I returned to NYU in early 1970, I was replaced as administrator by Jerome Kretchmer, a politician who, in one of his first actions, cancelled the contracts for the two incinerators on the grounds that they would be unacceptable sources of air pollution. Many years later, in 1985, the Navy Yard project was revitalized, and both Kretchmer and I were requested to appear before a city council committee which was again considering the matter in the face of opposition from the local community. I presented the reasons why I had originally supported the proposed construction, and why those convictions had been strengthened by developments in the intervening years. I was pleased that Kretchmer reversed his original position and supported my statement. However, by that time neighborhood opposition was even better organized than in 1969, and construction of that badly need incinerator has not begun as of this writing.

A crisis developed in New York City in February 1969 as a result of a snowstorm that paralyzed the city for days. On a Sunday when only flurries were forecast, a storm developed that dumped fifteen inches of snow on the weather observation station in Central Park, and twenty-two inches at Kennedy Airport. Since it is the responsibility of the Department of Sanitation to remove snow, a major storm results in the mobilization of all men and equipment. Refuse collection is suspended, plows are fitted to most of the collection trucks, and huge piles of salt maintained at strategic locations throughout the city are loaded into spreaders. The battle against snow is carried on in military fashion, controlled around the clock from the Department of Sanitation headquarters by the commissioner and his principal lieutenants.

Irma and I had decided to spend that weekend in Sterling Forest. When it began to snow on Sunday morning I discussed with Commissioner Moeller whether I should return to New York, but decided not to do so because of a favorable weather forecast. However, things deteriorated rapidly and by late afternoon we began the forty-mile drive to the city. I knew nothing about snow removal, but I was naturally interested in the manner in which the twenty thousand sanitation men were deployed in a snow emergency. However, I had been warned that in a serious snow emergency city hall received all kinds of inquiries and complaints from the public and that, as the responsible administrator, I was expected to deal with many of them.

I was driving the radio-equipped Buick assigned to me, and became aware that the intensity of the snowfall was increasing as we slowly approached New York City. As we came into the range of the sanitation truck communication system, I began to appreciate the drama that was in progress: The usual medical emergencies were complicated by the inability of ambulances to maneuver through clogged side streets. People were stranded in cars, and the parkways were being closed to traffic.

We barely made it into lower Manhattan, where I could see the problems first hand. Many intersections were blocked by abandoned cars, which were already encased in snow. More ominously, I found a surprisingly large number of Sanitation Department trucks apparently broken down and abandoned, some of them in the middle of streets. The curbs were cluttered with parked private cars that blocked the work of the snow plows and there was virtually no moving traffic. Before the snow stopped the next day, the city was beset with its most serious snow emergency in decades. The airports, railroads, and main highways were out of service for much of that week, and the life of the city was at a low ebb.

I did not realize the extent to which I would be involved personally in the recriminations that developed until a day or two later, when an editorial in the New York Times suggested that one of the reasons why the city was paralyzed by snow was that I was in my "country estate" when the storm arrived! This, it turned out, was an erroneous scoop given to the Times by one of the officials of the sanitationmen's union, which was lobbying to have the department made independent of the EPA.

The storm was officially recorded as a fifteen-inch snowfall, because the records are based on measurements made at the weather observation station in Central Park. This was bad enough for New York City, but a more appropriate figure would have been twenty-two inches, because that was the depth recorded at John F. Kennedy Airport. It was the borough of Queens, where the airports are located, that was most severely affected. The most recent storm of that severity had been in 1947, twenty-two years before, when conditions were almost identical, the city was caught unprepared, and several days of near paralysis also followed.

At the time of the storm a mayoral election campaign was just getting underway. Lindsay, who was seeking reelection, was attacked quickly, particularly by the Democratic organization in Queens. Moeller and I also came under attack. On February 18, when we were both exhausted by the events of the previous nine days, the city council conducted a hearing. It developed into a disorderly affair, which was not unusual for that legislative body. Commissioner Moeller had prepared an opening statement in which he had planned to analyze the reasons for the severe disruption in the life of the city, but he was not allowed to read it. The councilmen from Queens attacked us for an entire morning, without giving either of us a chance to reply. This was legislature behavior at its worst. I think it was both proper and necessary for the city council to conduct an inquiry, because they had a duty to understand the reasons why the costly disruption of the city had happened. But the end of the snow emergency had only been declared the day before, and all of the officials involved in the ordeal, including Moeller and myself were exhausted by the strain of round-the-clock duty for more than a week. It would have been a more productive hearing had we been given a chance to regain our strength and gather the necessary facts.

About two weeks later, after a five-inch snowfall had caused little inconvenience to the city, I was asked by a reporter to explain the reason for the improvement. Remembering that the previous storm had deposited twenty-two inches on Queens, I replied, "Seventeen inches!"

AIR POLLUTION CONTROL

By the time I took office in March 1968, the environmental movement was well underway in the United States, and most of the emphasis in the large cities was on air pollution. More than three thousand Londoners had died during a three-day smog in 1952, and that event was widely believed to herald even worse disasters in other large cities. When John Lindsay was elected mayor in 1965, one of his first actions was to appoint Norman Cousins, publisher of the *Saturday Review of Literature*, to head a Task Force on Air Pollution in the City of New York. The report of the task force, entitled *Freedom to Breathe,* was published in the spring of 1966. Its many well-considered recommendations were a blueprint for the enormous job to be done.

Among the findings of the Cousins report was that the city government was the worst violator of its own laws. There were eleven large municipal incinerators, which had forty-seven furnaces and smoke stacks with little or no pollution control equipment. The New York City Housing Authority operated more than twenty-five hundred incinerators and heating furnaces, most of which lacked air pollution control equipment. The city also operated an asphalt plant on the East Side of Manhattan that was a constant source of annoyance. Finally, many schools were heated by ancient coal-burning furnaces, and the engine exhausts from the thousands of city-owned buses were ignored.

Other major sources of pollution were the power stations operated by the Consolidated Edison Company, thousands of apartment house incinerators and heating plants, eighty-five hundred industrial establishments, nearly two million automobiles and trucks, hundreds of thousands of aircraft arrivals and departures, and the twenty-five thousand steamships that used New York harbor each year. In addition, demolition and construction, a major industry in New York City, were a major source of dust as well as smoke from burning waste in open fires.

Although the Cousins report had identified the sources of air pollution, I knew it was important to develop information about the relative importance of the sources of air pollution. Even before taking office, although I was already on leave from NYU, I succeeded in obtaining a research grant from the state of New York which would permit our laboratory to explore methods for apportioning the relative contributions of the various sources to the pollution of the city's air. Ted Kneip, with Michael Kleinman as his graduate assistant, tackled the problem with great effectiveness and provided the city with the basic technical information needed to assign priorities to the required control measures.

The approach they took can be illustrated by one example. The main fossil fuel being burned was oil, since a decision had been made to outlaw burning of coal within the city limits. When samples of airborne dust were collected daily over a one-year period and analyzed for various trace elements, it was found that the total dust concentration varied with temperature as well as with the vanadium content of the samples. Since vanadium is present in fuel oil, the clear implication of the finding was that space heating was the principal source of particulate pollution. Other "fingerprints" were identified that assisted us to allocate the importance of incinerators, automobiles, and wind blown dust.[1]

The Cousins report inspired a program of action that dramatically improved air quality during the next two years. Much of the credit goes to the city council, which incorporated many of the report's recommendations into a new air pollution control law (Local Law 14) that mandated certain basic requirements, among which were limits on the sulfur content of fuels, prohibition of on-site incineration in newly constructed buildings, shutdown or upgrading of existing incinerators, and elimination of open burning of garbage and leaves.

Credit for the improvements that took place also belongs to Austin Heller, who had been recruited as commissioner of air pollution control about two years before my arrival and assisted in drafting the law that evolved from the Cousins report. Unfortunately, Heller was often overzealous in his approach to air pollution control so that he lost favor with the mayor because of his frequent altercations with other members of the administration. One of the first questions Lindsay asked me was whether I wanted to retain him in the new post of commissioner of air resources. There was never any doubt that I did, and only once during the next two years did I find it necessary to override any of Heller's decisions. During that hectic period when the city was tormented by sit-ins, burnings, racial riots, and demonstrations of all kinds, representatives of the Hispanic community applied to the Fire Department for a permit to build a bonfire as part of a traditional religious observance. Since bonfires were outlawed by the air pollution control law, the fire commissioner consulted Heller about issuing a variance but he refused. The fire commissioner was concerned about the ethnic sensitivity of the matter and when he appealed to me, I gave permission to conduct the ceremony on an island in the East River. The benefit from permitting this important ethnic population to conduct its religious ceremony more than offset any harm from the smoke produced by the bonfire.

Air pollution at sufficiently high levels can have serious, even fatal, effects on people, but the health effects at levels of air pollution encountered from day to day were not understood in 1968 and in fact are not adequately understood even today. There were no air pollution standards to provide a basis for deciding how far to proceed in our program. Of course, many people would like to have the cleanest air possible, but air pollution abatement costs a good deal of money, and the costs increase exponentially as the standards become more strict. It was estimated in 1969 that about $500 million would be spent in New York City to achieve our clean air goals. At what point would further investment in clean air be unwise because the funds could be better spent on water pollution control, better housing for the poor, or any of the numerous ways in which the public health can be improved?

The question was particularly urgent with respect to sulfur oxides, which are produced by the burning of coal and oil. It became apparent that the limitations we had placed on the sulfur content of fuels used in New York City would result in a substantial reduction in the concentration of sulfur dioxide in the air of the city. In 1966 the average concentration of this noxious gas reached 0.22 parts per million (ppm), but by 1969 it had been substantially reduced, and was expected to average 0.06 ppm in 1970. How clean is clean enough? How far were we justified in demanding further reductions in the sulfur content of fossil fuels, knowing that the lower the sulfur content, the greater the cost?

In the absence of federal guidelines such as exist today, I asked the commissioner of health to assemble a group of the country's top experts for guidance. After reviewing the scanty epidemiological and laboratory evidence, they recommended that we set the limit at 0.06 ppm, the level we expected to achieve the following year. The experts also made the important recommendation that New York City initiate epidemiological studies so that the effects of exposure to low levels of sulfur dioxide could be determined. This seemed a sensible course of action, but we knew that it was necessary to consult the U.S. Department of Health Education and Welfare (HEW) which was developing a system of ambient air quality standards. (That function was later removed from HEW and assigned to the U.S. Environmental Protection Agency when it was formed in 1970.) When we consulted the responsible officials in Washington, they suggested that we should be more cautious and should reduce the proposed permissible concentration of sulfur dioxide by 50 percent to 0.03 ppm.

This was my first experience with the phenomenon of "ratcheting" in the development of environmental standards. HEW had no scientific justification for the reduction. But neither could they admit that anyone other than themselves knew what was adequately safe. Whatever figure we proposed would of necessity be too high by a factor of two. I have seen this happen over and over in the struggle that has developed among bureaucracies during the past twenty years to decide which is the cleanest of the clean. In this case, no one would have objected if there had been evidence that the reduction was necessary, but there was no such evidence, and our proposal was based on the best possible advice. It wouldn't have mattered if the change hadn't been so costly. We estimated that the annual operating costs of reducing the level from 0.22 ppm to 0.06 ppm would be $70 million per year but that it would cost $200 million per year to reach 0.03 ppm.

The amount of settled and airborne dust has been reduced enormously in many of our cities, including New York, as a result of the gradual replacement of coal for domestic heating and power generation. This has come about largely as a result of the availability of oil and gas, which have been much favored because of their cleanliness and convenience. Con Ed, to the credit of its management, voluntarily gave up the use of coal for power generation in favor of oil with a low sulfur content during the period when the city's EPA was developing its strategy for air pollution control. However, the largest sources of pollution were not the power plants, but thirty thousand apartment houses which burned a poor quality fuel oil for domestic heating. Although the smoke from the utility's high smokestacks was far more conspicuous, its overall effect was not as offensive as the thousands of smaller plumes of black smoke from the apartment houses. This resulted from the poor quality of the oil they used, as well as improper operation of the boiler equipment. The most difficult part of our cleanup campaign was obtaining compliance with the new law that required these boilers to be maintained and operated properly. Hundreds of apartment house janitors were required to attend classes organized by Commissioner Heller and his staff to teach them proper operating and maintenance procedures.

The second largest source of dust and soot was the seventeen hundred apartment house incinerators that, as noted earlier, were installed prior to 1967 at the insistence of the Department of Sanitation. In 1968, installation of new apartment house incinerators was banned, and the apartment house owners were required to improve the performance of the existing equipment.

The amount of money spent by the private and public sectors to clean the air of New York City was huge, but well worth while. People who live in a city with seven million inhabitants, two million automobiles, and hundreds of thousands of commuters cannot expect to enjoy the air quality found in many rural and suburban localities, but it can be said that New York, along with other cities such as London, St. Louis, Pittsburgh, and Chicago made enormous improvements during the late 1960s. I was glad to have been part of that successful effort.

NOISE ABATEMENT

Noise is unwanted sound, as judged by the listener. The person who finds it relaxing to listen to a Brahms symphony at midnight may be creating noise for someone who doesn't want to hear it.

The law that created the EPA specified that it should develop a noise reduction program, and I accordingly established a Bureau of Noise Abatement within the Department of Air Resources. As air pollution control was complicated by the tens of thousands of individual sources of pollution, noise abatement was even more complicated because the problem originated from hundreds of thousands of sources. Whereas the air pollutants tended to diffuse and contaminate the air of an entire neighborhood, the noise was usually more localized, perhaps affecting only a few people, as in the case of a roof-top fan located next to an apartment in an adjacent taller building. Or, the noise might come from a crying baby or an arguing couple. Every garbage can that resonated against the curb in the early morning hours was a source of noise.

Thus, the goal of reducing the noise of a large city was difficult to achieve. High on the lists of priorities had to be construction equipment (particularly air compressors), sanitation trucks, sirens, autos and trucks, and aircraft. A task force which included several acoustic experts was appointed to study the problem. Its report *Towards a Quieter City* served as a basis for getting started.

One of the most important sources of complaints was the Sanitation Department. Many refuse collections were in the early morning hours and disturbed people's sleep as the metal garbage cans in general use clanged against the pavement. The refuse was dumped noisily from the cans into a hopper at the rear of the trucks, after which a compactor with a loud whining noise pressed the garbage into the body of the truck. The garbage cans and the compactors were major acoustic nuisances, so a program was immediately begun to deal with them.

We recognized that the trucks could not be altered in the short term. They were expensive pieces of equipment and, while we could accelerate their retirement somewhat, the city could not afford to replace immediately the thousands of trucks it owned. However, the truck manufacturers came up with some innovative designs that substantially reduced the whine of the compactors, and these gradually replaced the older models. The noise generated by the metal garbage cans was greatly reduced by the introduction of plastic bags, as noted earlier.

Setting acoustic standards for equipment such as rooftop air conditioning equipment or air compressors takes a good deal of time. Research must be undertaken to provide information about redesigning the machines to produce less noise. If the equipment cannot be retrofitted, a period of time approximately equal to its expected life must be allowed before it is required to be replaced with the newer design. Otherwise, the cost of discarding tens of thousands of units is prohibitive. My involvement in efforts to do something about such sources of noise led to nothing but frustration. The city council as well as citizen action organizations wanted instant results, which was not possible. Even efforts to develop long-range plans came to nought. To develop standards of performance, we needed the cooperation of the companies that manufactured the machinery, but this could not be obtained because by seeking cooperation, we were encouraging violation of the antitrust laws.

WATER SUPPLY

New York City is blessed with a supply of excellent water carried in deep rock tunnels from watersheds as far away as 125 miles. The vast system built and maintained by the city provides water not only for its own needs, but also for eight upstate counties. The per capita demand for water had risen steadily from about twenty-five gallons per day in the early part of the nineteenth century, to more than 150 gallons per day in the late 1960s, when the supply was 1400 million gallons per day and was expected to increase to 2200 million gallons per day after the year 2000.

Water supply in New York has been the subject of financial machinations from the earliest days of the republic. None other than Aaron Burr arranged in 1799 for the water supply franchise to be given to a private company, with the stipulation that the company was allowed to invest its surplus capital in financial transactions. The Manhattan Company, as it was called, found its financial activities so profitable that it never became a serious supplier of water but evolved instead into the Chase-Manhattan Bank, one of the country's largest financial institutions. The company not only neglected its mandate to develop an adequate water system, but also used its political influence to keep others from doing so for more than forty years.

When a municipal water supply system was formed in 1835, it immediately became enmeshed in politics because it involved large construction projects and jobs that could be dispensed by politicians to thousands of the immigrants from Ireland who were pouring into New York. In 1899 the politicians again attempted to strike a deal to place the water supply in private hands under conditions that would have resulted in outrageous profits for the manipulators. In 1905, the New York state legislature created the Board of Water Supply of the City of New York, with the responsibility for planning and building the required water supply system. The board was given only planning and construction responsibility. Operation of the supply was assigned to the Department of Water Supply, Gas, and Electricity, where it remained until its Bureau of Water Supply was renamed the Department of Water Resources and absorbed into the EPA when it was formed in 1968.

As EPA administrator, I was by law an ex-officio member of the board. Shortly after I took office, Judge Edward McGuire resigned as president of the board, and I was elected to take his place. I served in that capacity for the next two years.

The Board of Water Supply was unusual in many respects. Its three members were appointed by the mayor and held tenure for life. To keep the board out of partisan politics, the legislature gave the board unusual independence of the normal checks and balances to which other city agencies are subject. It maintained a staff of four hundred engineers, who were greatly respected by their peers. The network of watersheds, aqueducts, and reservoirs they designed and built are engineering marvels. The huge projects they controlled were financed by bonds that were issued independently of the city's debt limit that applied to other needs, such as hospitals, schools, parks, and roads. Above all, the funds available to the board could be expended without the need for line-by-line approval of the city budget director, as was required of other city agencies. The unusual responsibilities and powers of the board had for many decades been administered in an exemplary way, entirely free of scandal. To avoid scandal was, of course, the underlying reason why the board was created in 1905.

I joined the board at a critical time in its history. The water supply reached the city through two deep rock tunnels that were built in the early part of the century. Neither tunnel had ever been shut down for in-spection and maintenance. By 1969 one tunnel carried 40 percent of the city's water supply, and the other 60 percent. If an emergency developed in either tunnel, the remaining capacity would not be sufficient to supply the city's water needs. Moreover, significant changes had taken place in the population distribution, with major shifts to some of the outlying areas. For this reason there were parts of the boroughs of Queens and Staten Island where the water pressure was not sufficient for fire fighting.

The board had decided in 1966 that the time had come to build a third water tunnel (usually called the third city tunnel), which promised to be the most expensive civilian construction project in history. When I ar-rived on the scene I found that the project had been stalled for two years because of opposition from the budget director and City Planning Com-mission. Both based their opposition on advice they were getting from a corps of bright young planners and systems analysts who flocked to gov-ernment at all levels in the 1960s and practiced the concept that any pro-posed project, however complex, could be solved by the application of logic, without the need for experience or conventional methods of engi-neering analysis and design. I have often found the analysts helpful in planning and executing complex programs, but their techniques can only be effective when combined with the work of the pragmatists.

Both Lindsay and Budget Director Fred Hayes, however, looked with favor on the young analysts and didn't relate well to the pragmatic and seasoned veterans who were the very heart of New York City govern-ment. Most of the engineers, such as Maurice Feldman, whom I chose to be commissioner of water resources, Vincent Terenzio, the chief engineer for the Board of Water Supply, and Martin Lang, director of the Bureau of Water Pollution Control, had been at the tops of their classes when they graduated from engineering school during the depression. In the 1930s a civil service position in New York City was very attractive, and they were appointed to entry level positions after taking tough examina-tions in competition with hundreds of other applicants. Over the years they gained worldwide recognition only to find themselves suddenly con-fronted by the new breed of systems analysts. In New York City at that particular time, the young systems analysts often did not work well with engineers.

In 1968, we were in the midst of a period of unrest among young people who were seeking social change. Beginning with public frustration about the war in Vietnam, public disquiet soon erupted over inequities in civil rights, restraints in sexual behavior, and environmental issues. A revolution was in progress that has changed society in many ways, not all of which have been for the good. One of the casualties of that period was the "expert." This resulted from the general distrust of authority in government, on the campus, and even in the home. The irreverent young analysts, fresh from the Ivy League schools with degrees in business administration or law, wanted to remake city government. They wanted the public to understand how decisions were made and to participate in the process of making them.

The need for the professional to communicate effectively with the public goes without saying. But interested members of the public must be willing to devote time to studying the issues and understanding the technical issues involved. This the public is often unwilling or unable to do. Today, twenty years later, the desire of the public to participate in decision making pervades our communities to an even greater degree than in 1969. There may be reasons why this is so. Watergate and the Iran-Contra affair are only two examples of matters in which the public believes its judgement was superior to that of the experts. I have been confronted with the distrust of experts many times since I first encountered it in the dispute over the third city water tunnel.

One conclusion I had reached during my period of orientation was that, of the many problems that came under the aegis of the EPA, the most urgent was the need to build the third city tunnel. The city could get by for a time without additional controls over air and water pollution. The streets were dirty, but this could be tolerated a while longer if necessary. But the city could not tolerate an interruption in its water supply.

I immediately found myself in the midst of the two-year-old confrontation between the board and the city's budget and planning agencies. The board had retained the nation's top engineering firms for advice. The Budget Bureau was leaning on information from its small group of bright, articulate planners. For about a year, the mayor found himself caught between his two sources of advice and decision making. The dispute had become bitter as time went on, particularly when one of the planners revealed his plan to inspect the two existing tunnels without the need to build a third: he suggested that the first and second tunnels be inspected from miniature submarines!

On taking office I explained to Lindsay that the need for the third tunnel was crucial in my opinion. I reminded him that the board had a good deal of independence and had the right to bypass the mayor and proceed directly to the Board of Estimate to obtain approval to build the tunnel. This, as president of the Board of Water Supply, I had the power to do. The issue was so important that I would not hesitate to take that route if the Budget Bureau continued to block the project based on objections that I believed lacked technical substance. Although I did not say so, it was obvious that this would be a political embarrassment to Lindsay, who was running for reelection in the election to be held in November 1969.

On October 9, one month before the election, the Budget Bureau withdrew its objections and the project proceeded. Until that point I had thought that the functions of the Board of Water Supply could be better performed if they were absorbed into the EPA. This was what Lindsay wanted to do because he was bothered by the independence of the board from city hall. However, my experience with the third city tunnel convinced me of the value of the independence granted to the board by the state legislature in 1905. Contrary to the wishes of Lindsay and his advisors, I fought to retain its independent status but it was abolished nevertheless soon after I left office and its functions turned over to the Department of Water Resources. I am sure that, for at least a while, the long range needs of the city for an abundant supply of water will be well managed. As for the long range requirements, I am not so sure.

There was one problem with the water supply system that I found particularly bothersome. Everyone who lives in a large city knows that there are frequent water main breaks that disrupt traffic, interrupt water supplies, and flood basements. Shortly after I assumed by duties, I inspected the consequences of a break in a forty-eight-inch main on Manhattan's East Side. A number of questions occurred to me: (1) how many breaks occurred per year, and was the number increasing with time; (2) what caused them; (3) how much did they cost, and (4) should there be a long-term program to reduce their frequency.

This was the first of many occasions when I realized that New York City lacked the basic statistics necessary for its proper management. The breaks were assumed to result from the inevitable effects of aging, but apparently no one had ever investigated whether the aging was due to metallurgical changes in the cast iron from which the pipes were made, or simply mechanical stress. If the latter, perhaps the effects could be reduced by laying the pipes in some different way, such as by placing a bed of sand in the trenches. The department had never consulted a metallurgist for guidance. Moreover, there were no charts showing the frequency with which breaks were occurring. Worst of all, there were no records of the costs of the breaks which, in many cases where buildings were flooded, must have been considerable.

On further investigation I concluded that the basic problem was that the budgetary constraints on the department simply discouraged innovation. The Budget Bureau kept the department on slim rations, which was penny-wise but pound-foolish. It occurred to me that the Department of Water Resources should be run as a municipally-owned utility. It should finance its operations by revenue bonds and amortize the bonds from income. The commissioner should have incentives to reduce costs so that the savings could be invested in innovative methods of both supplying water and treating wastewater. Under the existing method of operation, any money saved was simply returned to the general treasury. If operated as a utility, the department would be expected to make a "profit" that would be for the general benefit of the treasury. The main advantage would be that the department could be run in a businesslike way.

I had been approached by Webster H. Wilson, who had just retired as Chairman of the Board of the Hazeltine Corporation and who wanted to work with me to "help improve the city." I assigned him the task of investigating the advantages and disadvantages of operating the Department of Water Resources as a municipally owned utility. He produced a splendid report that showed that the best operated water supply systems in the country were in fact operated in this way and he recommended that New York City change its mode of operation accordingly. His report has been gathering dust somewhere in city hall for the past twenty years and would make useful reading for city officials, even now. But that is too much to expect. Like all too many large organizations, New York City government has little or no institutional memory.

WASTE WATER TREATMENT

For reasons that I never understood, water supply and waste water treatment were historically the responsibility of different departments. Water supply was in the Department of Water Supply, Gas, and Electricity, and waste water treatment was the responsibility of the Public Works Department. However, the two functions were similar in many respects and were combined into a single Department of Water Resources when I arrived.

During my two-year term of office we launched a $1.2 billion program of waste water treatment. The largest plant was intended to treat about 200 million gallons per day of sewage produced by residents on the west side of Manhattan. That massive amount of pollution was flowing untreated into the Hudson River. The plant, to be called the North River Wastewater Treatment Plant, had a controversial beginning that took a great deal of my time.

The original plan called for the facility to be located at about 56th street on the Hudson River. It was designed to provide for 67% removal of the biochemical oxygen demand (BOD), which is a measure of the amount of reduction in the requirement for oxidation of organic materials in the effluent after it is discharged to the receiving water. The higher the BOD removal, the less is the reduction in the dissolved oxygen content of the river water. The required amount of BOD removal is determined by the volume of the receiving body of water, the quantity of organic material in the effluent, the rate of tidal exchange with the ocean, and the concentration of dissolved oxygen content in the receiving water. One of the reasons for treating sewage is to maintain the concentration of oxygen in the receiving water above the level required for its ecological health. The planned 67% BOD removal was thought sufficient, considering the physical and chemical factors that existed. A contract for the design of the plant had been completed in 1966, and it was estimated that its construction would cost about $250 million.

The contract would have been let before I took office had it not been for a combination of circumstances that arose. First and foremost was the fact that the environmental movement of the mid-1960s was in full swing, and there was a clamor for "clean water." This was for the good: it was really disgraceful that 200 million gallons of raw sewage were flowing into the Hudson River each day. However, how clean is clean enough? There was no answer to this question, except that in the judgement of the public health engineers, 67% BOD removal was sufficient in the case of this particular plant. However, two men who were running for office thought New York should have the cleanest water money can buy, and campaigned on the promise that they would upgrade the performance of the plant to 90% BOD removal. The two men were Robert Kennedy, running for the U.S. Senate, and Fitz Ryan, running for reelection as congressman from a district on the West Side of Manhattan. Both candidates were elected, and the federal government soon let it be known that the city would not be eligible for financial construction assistance unless the plant was upgraded. When the plant was redesigned, we learned that its cost had increased from $250 million to more than $700 million, in 1969 dollars.

Another problem created by this decision was that the upgraded plant would be so large that it could not fit on the site originally chosen. Alternatives were few on Manhattan's West Side, and the best choice seemed to be a tract along the river's edge at 135th Street in West Harlem. Unfortunately, the decision was made at a time when that part of Manhattan was in a foment over an unrelated matter, construction of the Columbia University gymnasium, a subject that had racial overtones, and was the cause of prolonged violence. When it was announced that one of the largest sewage treatment plants in the world would be built just north of that neighborhood, the residents were infuriated.

It was clear that I needed a community affairs representative on my staff, and I was fortunate in attracting George Gregory, who had been the first of the great black basketball players when he played on the Columbia team in the 1930s. Gregory and I spent many evenings meeting with Harlem community leaders, and we finally obtained their agreement to allow the plant to be built. This resulted less from our power of persuasion, than from the fact that we were able to promise that the plant would be covered by a $50 million deck on which a new park would be constructed. Riverside Park, which runs along the Hudson for much of the length of Manhattan is so constricted at 135th Street that it almost disappears. It was planned that the plant would extend into the river, and the idea of the park located over the water was attractive to the residents.

What should the park include? To assist in this decision, August Heckscher, commissioner of parks, suggested that we retain Philip Johnson, the most prominent architect in the city. (Johnson was known for his ingenious use of fountains in his designs of skyscrapers then being constructed on Manhattan.) In due time he submitted plans for a park that would be covered by beautiful fountains. This was unacceptable to the residents who wanted maximum use of the deck for recreational purposes, not fountains. Gregory, to his credit, suggested that we retain an architect from the Harlem community. This we did, and he proceeded to meet with the residents to learn about their preferences. Not surprisingly, they wanted a swimming pool, tennis courts, and other amenities to satisfy their need for recreational facilities.

It is now about twenty years since construction of the plant was started. Work stopped when the city found that the cost was escalating during the inflationary period of the late 1970s, and the design was changed so that the BOD removal will be close to what was originally thought necessary. The two politicians who were misled by well meaning but uninformed advisors caused the city to spend hundreds of millions of dollars unnecessarily. In the meantime, raw sewage has flowed into the river for many years after the original plant could have been completed.

"THE MOST COMPETENT REPLACEMENT"

Lindsay was repudiated by both the Republican and Democratic organizations during the 1969 campaign, but ran as an independent and was reelected in November for a second term after a fierce contest. Shortly after his reelection I reminded him that my leave from NYU would end in mid-March and that a replacement for me would be needed. We both agreed that we should seek a new administrator from outside city government, and he asked me to take the lead in conducting the search. He also asked others to suggest possibilities, and Norman Cousins in particular became active in the recruitment. My only instructions were to search the country and recommend the most competent replacement I could find.

Several nationally known engineers and scientists were approached, and one of them, James Fitzpatrick, seemed especially well qualified. He was an engineer with a talent for administration, who had spent most of his career in the administration of Richard Daly, the mayor of Chicago. He had held several high posts, the most relevant of which was as commissioner of air pollution control. In 1970 he was living in New York, where he was president of a firm that manufactured air pollution control equipment. I had met him only briefly on a few occasions, but I was aware that he found his employment in the private sector dull and was anxious to return to public service. I approached him about the possibility of replacing me and was delighted that he expressed keen interest. Lindsay took a liking to him, and it was agreed by year-end that he would be appointed to replace me when my term of office ended in mid-March. Accordingly, in preparation for his new assignment, Fitzpatrick resigned his position and went on a winter vacation in Florida.

Fitzpatrick was still on vacation in mid-January, when I received a phone call from Lindsay early one Sunday evening asking if I could meet with him at Gracie Mansion as soon as possible. Irma and I were just about to leave our apartment for the Metropolitan Opera, but Lindsay rarely called personally when he wanted to see me and I knew that he must have something urgent to talk about. I told him I would see him within the hour and left immediately, proceeding first to the Met, where I dropped Irma, knowing that I would be able to return later in the evening.

Lindsay was in his small office in the basement when I reached the mansion. He seemed distracted and ill at ease, which I detected because he welcomed me leaning against a fireplace mantle, rather than from the comfortable chairs he normally used when we met. He abruptly told me that he was going to appoint "Kretchmer" as my successor. I recalled that earlier in the day I had had a call from a *Times* reporter asking me about a rumor that Kretchmer would be appointed. I had replied to the reporter as I did to the mayor, "Who the hell is he?" Lindsay replied that he was a former member of the New York State Assembly from the lower west side of Manhattan and a lawyer. I had the feeling that I must be dreaming: it was only two weeks earlier that Lindsay had congratulated me for finding Fitzpatrick, the highly qualified professional he was looking for, and to whom he had offered the position of EPA administrator upon my return to New York University in two months. Lindsay didn't offer any explanation for the sudden change, and I could see, after some prodding that I wasn't about to receive any. When I saw that his mind was made up I said simply, "You are making a great mistake, John". He did not reply.

The mayor understood the position I was in with respect to Fitzpatrick and offered to call him in Florida to spare me the pain of having to do so. He had obviously anticipated that this would be necessary, because he had Fitzpatrick's Florida telephone number in his pocket. I have never understood why Lindsay changed his mind. I have been told by others that Kretchmer was active in the liberal wing of the Democratic Party, and that Lindsay had incurred some obligations because of that faction's support for him during the recent election. I don't understand such things and still cannot believe they happen in real life.

The appointment of Kretchmer was announced next day and I was delighted that the *Times* reporter quoted my reply, "Who the hell is he?" in his account of the matter. I was glad to have it known that I had no part in Kretchmer's selection. The newspaper comments were generally critical of the appointment and a number of public figures made uncomplimentary statements about the absurdity of filling one of the most technical of the super cabinet positions with a politician who had neither the required training nor experience.

The mayor recognized that there was nothing in Kretchmer's background to qualify him for the position, and many of his close associates were quietly critical. One of them, who had served first as his executive assistant and had then moved up to become a commissioner of one of the more important departments, told me he was appalled by the appointment, and proved it by resigning from the administration. Many people inside and outside of the city government were shocked that the mayor had suddenly returned to a policy of filling key positions in the super cabinet with political associates rather than professional specialists.

Lindsay wanted me to work closely with Kretchmer to assure an orderly transition, and with two more months before he assumed office I agreed to do so. However, my relationship with Kretchmer got off to a poor start and deteriorated with each passing day. A few days after his selection was announced, he called me at home one evening wanting to know about some complaints from Staten Island residents about an annoying odor that had been persisting for the past few hours. This was not an unusual occurrence in that part of the city, where there was a considerable concentration of chemical industry facilities. I told him that we were aware of the complaints and that the matter was under investigation. He told me he was about to make a statement to the press to which I responded that he was free to do so as a private citizen, but I preferred that he not speak for the EPA until he assumed office. He didn't seem to understand the difference between being administrator and administrator-designate.

Far more serious difficulties developed during the next few weeks. I invited Kretchmer to attend a session of the Board of Estimate, where I was scheduled to present the details of a billion dollar budget for capital programs, including incinerators, sewage treatment plants, additions to the fleet of sanitation trucks, and the third city water tunnel. I thought it important that he be present at the hearing, which was the most important of the entire fiscal year. He appeared late, and only stayed for a few minutes. Later, I instructed the commissioners to prepare briefing materials for his orientation, but I was never able to arrange the series of briefings that should normally occur as part of the transition that was in process.

It became rapidly apparent that Kretchmer was demoralizing the EPA personnel at all levels. The three EPA commissioners indicated their wish to retire. Kretchmer's proposed staff, young, inexperienced, and impolite, began to come into our offices each day, announcing their plans to anyone who would listen, but introducing themselves to nobody. I came into my office one morning to find that during the night somebody had been examining my files. I explained to Deputy Mayor Dick Aurelio that it was essential that he keep the administrator-designate under control for a few weeks, or that it would be necessary for me to leave immediately to avoid the disruptive confusion that was developing. He found the latter option preferable, and I left my position in the Lindsay administration six weeks ahead of schedule. Lindsay sent me a flattering letter of appreciation, which he released to the press, but it was not a pleasant parting. When I joined his administration nearly two years earlier I thought I was associated with a leader who was destined to rise to even greater heights in national politics. Lindsay was frequently mentioned in the press as being a potential candidate for presidential office. When I left him, the structure that he had created painstakingly during his first four years was in the process of crumbling.

When my departure was announced, I received a call from the editor of *Science,* requesting that I prepare an article for early publication in that journal. I accepted, and assembled a straightforward summary of what I had learned during my two years in office. As I read it now, nearly twenty years later, I find one paragraph that summarizes the problems faced by an environmental protection administrator.

> ... deficiencies in the political apparatus of communities have traditionally frustrated an orderly solution to complex problems, and it is to be hoped that this factor will not be an impediment to effective environmental rehabilitation. The elected officials, the bureaucracy of government, and the newspapers are important components of the social substrate from which all governmental programs must be developed and nourished. Professional environmental health specialists can define the objectives, develop the timetables, estimate the costs, and, as we have seen earlier, be given substantial amounts of money with which to do the job. But factors that are related to the individual components of the political apparatus frequently cause issues to arise that seem extraneous to the job that must be done. The original objectives are sometimes overlooked, and priorities become misaligned. An important function of government is to permit the development of

thoroughly considered plans of action that can be implemented by professional leaders who are given authority commensurate with the responsibilities assigned to them. A community that allows itself to fail in these respects will be unable to deal with the ecological problems that face it.[2]

The EPA established by John Lindsay was the first of its kind, preceding by about three years the establishment of the large and powerful EPA in the federal government. Unhappily, the EPA in new York City was downgraded before long by removal of the Sanitation Department. Then all super agencies were eliminated, and the EPA became the Department of Environmental Protection. The fine team of professionals that I assembled left office shortly after I did. Kretchmer appointed himself commissioner of sanitation while still serving as administrator and, mindful of the importance of clean streets to the people of New York, let it be known that he would be riding into city hall on a garbage wagon. He made no secret of his political ambitions. For a while he was popular with the young activists who dominated the environmental movement in the early 1970s, but he soon returned to the practice of law and has not since been associated with environmental protection in any way of which I am aware.

James Fitzpatrick was remarkably understanding about the matter. He returned from his Florida vacation without a job, but with no apparent bitterness towards either the mayor or myself.

During the few weeks that had transpired since my retirement was announced, a considerable number of offers had come to me from industrial companies in need of advice about environmental matters. The National Environmental Protection Act had been passed overwhelmingly by the Congress, and one of the requirements was that environmental impact statements must be prepared as part of the procedure for licensing construction of new industrial facilities. I was well known to the executives of many companies, particularly the electrical utilities, and because there were few advisors in the field, I was receiving requests to provide guidance in the writing of environmental impact statements as well as for assistance with other matters related to environmental protection. Since Jim Fitzpatrick was without a job partly as a result of my abortive attempt to recruit him as my successor, it occurred to me that I should refer some of the requests to him. Within a few weeks he had a backlog sufficient to keep him busy for the foreseeable future, and he suggested that we form a consulting partnership. This seemed like an exciting opportunity, but I was expected back at NYU, from which I had been on leave for two years. Until then, it had never occurred to me that I would not return, but suddenly the possibility of starting yet another career was tempting, not only because of the challenge it offered, but because of the financial opportunities as well. Irma and I had lived comfortably on my academic salary, which had been augmented by occasional arrangements for private consultation, but in 1970 our three boys were finishing their educations (medicine, business administration, and law), and we found it necessary to dip deeply into our financial reserves. Further depletion had occurred during the two years in New York City because of the need to maintain two domiciles. I discussed the matter with Norton Nelson to determine if he would agree to my devoting half time to NYU, with the understanding that I would devote the remaining time to the venture with Fitzpatrick. He agreed that this was a workable arrangement.

Fitzpatrick and I formed a company, Environmental Analysts Inc., which quickly developed into one of the first of the successful consulting groups in the field of environmental protection. In 1975 we sold EAI, as our company came to be known, to the Equitable Life Assurance Company. I returned full-time to NYU, where I continued my career for another decade before retiring as professor emeritus in 1985 when I reached my seventieth birthday.

People often ask whether I regretted having spent those two years with John Lindsay. I always answer never. The work was demanding, the public was unappreciative of the many good things we did, Lindsay and his political circle did not always use good judgement, and the politicians on the city council were ruthless. But it was an enriching experience. My understanding of the environment was greatly broadened, and we were the first to demonstrate the advantages of unifying the many environmental functions of government.

While I did not regret the two-year experience, I have been greatly disappointed at how little of the information we gained was transferred to Kretchmer and his successors. It was a case of total institutional amnesia. From time to time in later years when I encountered senior officials and inquired about the status of some recommendation we made or report we wrote, I received blank stares. They had no knowledge of what we did a few years before!

Did I ever miss the job? Only in one respect. For two years I was assigned a Buick and chauffeur that were available for my use at any time of day or night. This was a perquisite that was justified for some of the commissioners and administrators by the demanding schedules we were required to maintain, and the frequency with which we had to move quickly from one part of the city to another in response to emergencies. I learned to enjoy having the car and chauffeur waiting at curbside wherever I might be. I have missed no other feature of my life as an official, but the car and chauffeur were an amenity too easy to enjoy and too hard to forget.

NINE

Meeting, Writing, and Advising

ATTENDANCE AT PROFESSIONAL conferences, preparing articles and reports for publication, and serving in various advisory capacities has accounted for a considerable portion of my time over the years. These activities are essential to the exchange of new information and ideas, and are particularly important to scientists and engineers who work in fields that are developing rapidly. In addition, because my work was sometimes involved with the setting of public policy and the establishment of rules and regulations, I have been frequently called upon to serve on boards, committees, and panels that provide advice to industry and government.

My initiation into the affairs of the professional societies began inauspiciously with an invitation to serve as a member of the program committee for the annual meeting of the Metropolitan New York Safety Council. To the best of my recollection that was in 1941, when I was young enough to feel very important when I was asked to arrange a panel on occupational health. This being my first such assignment, I took it very seriously and visited some of the leading occupational health specialists in New York City to obtain their suggestions for participants. I was young and unknown, but I was received courteously by all and with their guidance was able to assemble what was an excellent panel. This I can say with confidence because I soon received many other requests to arrange similar programs!

There were very few industrial hygienists at that time, and when the American Industrial Hygiene Association (AIHA) was formed in 1939, a local section came into being in New York City. There were only eight or ten of us in the group, and we met once each month in a back room of a French restaurant near the present location of Lincoln Center. (It was symptomatic of that post-prohibition era that after a year or so we moved to a new location because one of our more prominent members objected to the existence of a bar!) Many of the members of our small group became leaders in the field of environmental health. This was the start of decades of association with men and women of experience who were always ready to share their knowledge with me.

The AIHA originally served the needs of all the industrial health subspecialties, including engineering, medicine, chemistry, and toxicology, but each group of specialists soon formed its own organization. The Health Physics Society, which was formed in 1955, was of particular importance to me because of my interest in radiological health. Another society in which I was active was the American Nuclear Society, which became a huge organization that served the needs of the physicists, engineers, chemists, and metallurgists involved in the nuclear sciences. I served on the boards of directors of all three societies, and was elected president of the Health Physics Society in 1965.

The annual conventions of the professional societies were at first the only opportunity to meet others in my field. Later, special conferences on a national and international level that dealt with specific subjects in much greater depth became popular. Many of these conferences could be attended only by invitation, and some required the preparation of a technical presentation.

Not all such conferences have been productive. But even though many plowed old ground, it was not always easy to decline participation. Travel expenses were always paid, and there would sometimes be an honorarium sufficient to permit Irma to accompany me. Moreover, the meetings often took place in fine resort areas in both the United States and abroad, so that much of my foreign travel has been the result of invitations to technical conferences. It is a part of human nature to want to be seen and heard at such conferences, however unpromising the advance program may seem. Even so, I am glad to say that I learned to say no to many invitations and never became part of the scientist jet set that was present at all conferences, particularly the international ones.

Scientific Writing

Since 1942 when my first article, "The Principal Health Hazards in Metal Finishing Departments, and Their Control," was published in an undistinguished trade magazine called *Metal Finishing,* I have published nearly two hundred articles, book chapters, and monographs.[1] Many of the articles were research reports coauthored with graduate students or other colleagues and published in either scientific journals or the proceedings of symposia. Some, such as my first article, were intended to be educational and were written for the layman and non-specialists.[2]

I like to write. It is relaxing and seems to be the best way for me to clarify my thinking. When I become involved with a new subject that requires me to read many scientific papers, I often find it difficult to fit the pieces together until I have summarized what I have learned in writing.

By the time I joined the NYU full-time faculty in 1959, I had been involved for more than ten years in studies of the sources of environmental radioactivity and the mechanisms by which radioactive substances reach humans after being introduced into soil, air, or water. These pathways can sometimes be very complicated and involve subtle ecological relationships. I planned to call my first new course "Environmental Radioactivity," but there was hardly any written material on the subject. In contrast, the biological effects of radiation had been well studied, so that there were already textbooks on the subject. There was also an ample body of literature on the physical properties of the radiations, and the methods by which they could be measured. However, in 1959 the only writings on the subject of the ecological behavior of radioactive substances in the environment were a few government reports, the first reports of the United Nations Committee on the Effects of Atomic Radiation, and a few journal articles. To fill the need for a text book I began to write one—*Environmental Radioactivity,* which was first published in 1963 and is now in its third edition.[3] The book has been successful both as a textbook and as a standard reference work on the shelves of major libraries, and has been translated into Russian and Japanese.

In the mid-1970s, I became concerned that the subject of environment was becoming distorted in the public mind, and that environmental protection priorities had become misaligned. In response, I wrote a book for the layman entitled *Environment, Technology, and Health: Human Ecology in Historic Perspective.*[4] It received excellent reviews, and many complementary letters were sent to me by legislators, business leaders, and even a few environmentalists, despite the fact that the book was iconoclastic and took issue with the positions some had taken on a number of major issues. However, its value was short-lived and it was soon out of print. After ten years I rarely see references to it any longer, but it served a useful purpose at the time.

Committees, Boards, and Panels

Participation on committees (including boards, panels, and "commissions") established by local, state, national, and international private and public organizations can be useful and rewarding, but also time-consuming and frustrating. I have no idea as to the total number of such assignments I have completed, but they must total at least fifty.

From the early 1950s until I moved to North Carolina in 1985, I served five New York governors in a number of committee assignments that varied from a seven-year appointment to the New York State Health Advisory Council to the chairmanship of ad hoc committees concerned with various environmental problems that arose from time to time. These included the high levels of mercury in fish from lakes in the Adirondack Mountains and the presence of PCBs in striped bass due to contamination of the Hudson River.

My service on the Health Advisory Council coincided with two related private-sector activities that also dealt with the health care delivery system in New York State. For nearly ten years I was on the Board of Managers of the State Community Aid Association (SCAA), and from 1969 until 1975 I served as a director of the Blue Cross and Blue Shield of Greater New York. My association with the three organizations occurred when the delivery of health care and other social services was becoming increasingly complicated. It was the start of the current period in which the costs of health services have been rapidly increasing, when there have been demands for increased health care, when there was the first realization of the need to accommodate the requirements of an aging population, when dependence on drugs was becoming more prevalent, and when new health problems were being created by changes in sexual mores (although we did not yet know about AIDS). It was also a period when there was an accelerating movement of middle-class New Yorkers to the suburbs and an inflow into the city of Blacks and Hispanics whose state of impoverishment increased demands for health care services.

My service with these organizations gave me insight into the intricacies of the health care system. I was impressed with the enormous problems faced by Blue Cross-Blue Shield, a much criticized system of health care insurance which was (and still is) at a hopeless disadvantage because of the absence of the checks and balances needed to control the costs of medical care. Every person wants the best medical care that can be provided, regardless of cost, and private physicians are all too willing to provide the care without consideration of cost-effectiveness. In fact, many people find it repugnant that patients should be denied services that are too costly in relation to the expected benefit. In addition, the vulnerability of the physician to malpractice lawsuits has encouraged cost-ineffective procedures and prescriptions. As a result, for more than twenty years the cost of medical care has been rising more rapidly than other components of the cost-of-living index. This trend will not be reversed without drastic changes in the health care system, and the legal and ethical framework within which it operates.

Advisor to Federal Agencies

Every federal agency has advisory committees that assist in the formulation of policy, monitor ongoing programs, and give advice on special matters. Service on these committees can demand considerable time and energy, but the committee members benefit by gaining insights that are helpful in their work.

Some of the advice the agencies require about environmental matters is provided by the National Research Council (NRC), and the National Council on Radiation Protection and Measurements (NCRP). The NRC is the operating arm of two organizations, the National Academy of Sciences (NAS), and the National Academy of Engineering (NAE). The NAS was chartered by the Congress in 1863 during the administration of Abraham Lincoln to provide advice to the federal government in matters of science and technology. The National Academy of Engineering was formed under the NAS charter in 1964. A third organization, the Institute of Medicine (IOM) was established under the NAS charter in 1970. While IOM is not formally part of the NRC, its activities are subject to many of its rules.

The NCRP had its origin in 1929 with establishment of the Advisory Committee on X-ray and Radium Protection. With the discovery of fission and the development of nuclear energy, the scope of the original committee was greatly expanded and its name was changed to NCRP. In 1964 the organization was granted a congressional charter.

I have been a member of NCRP since 1967 and was elected to honorary life membership in 1983. I was elected to membership in the national Academy of Engineering in 1977. The work of these organizations is performed through committees, whose members are selected for their expertise and serve without remuneration. The NCRP maintains about eighty scientific committees, and there are several hundred committees at work within the National Research Council. It is not necessary to be a member of either the academies or the NCRP to serve on the committees they establish.

These committees often deal with questions that have enormous financial as well as scientific implications. For example, in 1982 the Department of Energy asked NCRP to evaluate the criteria being used to determine the disposal procedures in use for wastes that contain traces of plutonium. The rule specified that if the waste contained less than 10 nanocuries of plutonium per gram of waste, it could be disposed of at government-owned shallow land burial facilities. At higher concentrations, the wastes were required to be emplaced in deep geological repositories, at far greater expense. The problem was that no one seemed to know the rationale for the 10 nanocuries per gram, so the NCRP was asked to examine the practices in use and recommend possible changes. I chaired the committee established for that purpose, which found that the amount of plutonium disposed of by shallow land burial could be increased to 100 nanocuries per gram with no additional risk. This reduced the need for expensive disposal in deep geological repositories to the extent that DOE saved $260 million during the first year our recommendation was implemented.

Over the years I have also served on a number of committees of other federal agencies, including the U.S. Public Health Service, Atomic Energy Commission, Environmental Protection Agency, Department of Energy, Nuclear Regulatory Commission, and NASA. Many of these assignments were very time consuming, but this was more than offset by the months of association with specialists from many fields.

Consultant and Advisor to Private Industry

I have been an individual consultant to private industry, as well as a member of advisory committees. Three such committees have been particularly notable. Two were advisory councils of major nonprofit organizations established by the electric utility industry, the Electric Power Research Institute (EPRI) and the Institute of Power Operations (INPO). The third was the Safety Advisory Board (SAB), established by the General Public Utilities Corporation to oversee the cleanup at Three Mile Island. I have from time to time also been a member of other committees in the private sector, mostly in the electric utility industry.

EPRI was established in the early 1970s at a time when the utilities were the target of frequent criticism because of the environmental effects of power plant effluents and their general lack of technological innovation. The latter issue was one of the main conclusions that emerged from investigations of a 1967 blackout of the entire northeastern part of the U.S. which revealed basic weaknesses in the plans of the utilities to ensure the supply of energy during emergencies. Although the utility industry was the largest in the U.S., it was doing very little research.

The EPRI Advisory Council consisted of scientists, engineers, economists, sociologists, state utility commissioners, heads of church and civic organizations, and industrial executives. The council did not become concerned with the internal functions of EPRI, but focussed on the intermediate and long-range problems of the utility industry. Since the utilities are so intimately involved with the economic health of the country, the meetings of the council often dealt with matters of basic national concern. In addition to quarterly two-day meetings, there were week-long retreats each summer to permit in-depth discussions of selected topics. The eight years I served on the EPRI Council provided me with a marvelous education, but I always had the feeling that I was getting more out of the meetings than I was contributing.

INPO was formed by the electric utilities in the wake of the 1979 accident at Three Mile Island, to promote improved safety and reliability in the operation of the nation's nuclear power plants. The Nuclear Regulatory Commission has the statutory responsibility for monitoring the safety of nuclear energy, but INPO was established to encourage the utilities to operate the plants at the highest possible level of safety and efficiency. In principle, INPO was the self-policing arm of the utility industry. Its advisory council was structured much like that of EPRI, but the meetings were more concerned with the operations of the organization and the progress it was making. A considerable improvement in the operating performance of the nuclear power plants in this country has occurred during the past ten years, due in a large measure to the effects of INPO. Not only has the danger of serious accidents been lowered, but there are fewer plant shutdowns and they are of shorter duration than formerly.

Of all the industrial advisory committees on which I have served, the Safety Advisory Board for the Three Mile Island cleanup was in many ways the most interesting, satisfying, and frustrating. The board consisted of experts from ten fields including risk assessment, medical radiology, reactor engineering, nuclear physics, political science, environmental science, health physics, and chemical engineering. The TMI cleanup was an enormous engineering undertaking that cost one billion dollars. The engineering problems were difficult; nothing like it had ever been done before. When the cleanup started, there was no information on conditions in the reactor building or the condition of the core. But step by step, the work proceeded with great care and ingenuity for more than ten years, during which time the crippled power plant was so carefully dismantled that the radiation exposures to the employees were well within the established limits, and there was no exposure of residents living near the plant.

The most difficult problems in the aftermath of the accident had to do with, not the engineering difficulties, but with community misperceptions of the risks to which local residents were exposed. Although the accident at TMI was catastrophic in the sense that a multibillion-dollar nuclear power plant was destroyed, the radioactivity released from the core of the reactor was almost entirely contained within the power plant buildings. The highest exposure to any resident near the plant was less than that received from natural sources of radiation in one year. The average exposure to nearby residents was about 8 mrem, which is well within the variations in the dose we receive from nature due to differences in altitude or the composition of the rocks and soils. For example, there is a 15 mrem difference in the dose received annually in Manhattan and that received in the borough of Queens, across the East River. The radiation exposure is higher in Manhattan because it is built on igneous rock, whereas Queens is built on sandy soil that has a smaller amount of natural radioactivity.

Although the radiation exposure of the nearby residents was insignificant, they endured a frightening experience. At the time of the accident, more than four hundred reporters and photographers descended on the small communities near Three Mile Island. For days, the printed and electronic media carried sensationalized reports of what was happening. Getting reliable information to the public was made difficult by the confusion that existed among various agencies of the government, and by the exaggerated statements made by a few antinuclear scientists who came to the scene. Some of these scientists, without foundation, stated that the government had deceived the public and that the actual radiation exposures were much higher than stated. Under these circumstances, it is understandable that some residents who experienced problems with family health following the accident could believe that the problems were due to radiation exposure. Under normal circumstances, a fraction of the population will be experiencing health problems at any given time, with or without radiation exposure. Women have miscarriages, children are born with defects, and people develop cancer. In the aftermath of Three Mile Island, lawyers lost no time in capitalizing on the misperceptions concerning the consequences of the accident: more than two thousand law suits brought against the utility will take years to adjudicate.

I have found that my consulting activities improved my effectiveness as a teacher and research director. I could better teach about air pollution control because I had assisted companies in solving difficult problems. I could talk firsthand about the problems of implementing a state water pollution control program because I had been a consultant to the state of New York. The accident at Three Mile Island was not something I had read about—I had been there. My consultations made it possible to obtain valuable "hands-on" experience in "the real world."

In the course of consultations I sometimes became aware of unexplored research areas and found it possible to initiate investigations that strengthened the program of my laboratory at New York University. Our program of ecological research on the Hudson River was started in 1962 because I became aware, as a Con Ed consultant, that information would be needed about the ecological characteristics of the river to make intelligent estimates of the environmental impact of nuclear power plants. I also realized that the hazard from radioactive iodine could not be evaluated properly without more information about the physical and chemical form in which it was released from fuel. Con Ed provided the means to build a small laboratory inside the containment building of Unit 1, where one of our graduate students did his doctoral research on the subject. After the accident at Three Mile Island, I became aware of the need for an instrument that could pinpoint sources of radiation in an extremely high-level radiation environment. The Electric Power Research Institute provided a grant to develop such a device. There were many such examples.

However, consulting presents the constant temptation of becoming overextended, which can easily result in loss of effectiveness both as a teacher and a consultant. This is one of the most important tests that a professor must pass, not only with respect to consulting activities, but also committee assignments and invitations to conferences. The inability to say no is a major cause of casualties among able and energetic academics.

Low-Level Radioactive Waste Management

In 1987, North Carolina Governor James B. Martin appointed me to be the first chairman of the newly formed fifteen-member North Carolina Low-Level Radioactive Waste Management Authority (LLRWMA). The authority was charged with finding a suitable site for, building, and operating a low-level radioactive waste repository to serve eight southeastern states.

Six commercially operated low-level repositories have been in operation in the United States, and all of them have generated controversy. By the early 1980s three of them had shut down. Continued operation of the remaining three was in doubt. In 1980 Congress decided that intervention by the federal government was necessary and passed the Low-Level Waste Management Policy Act which, among other things, encouraged the states to make regional arrangements for low-level radioactive waste disposal. A Southeast Compact was established as a result of this legislation and it was planned that each of the eight states should operate a facility for the benefit of the others for twenty years, at the end of which time another host state would be selected. North Carolina was selected as the location of the facility that would accommodate the needs of the Southeast Compact for twenty years beginning January 1, 1993.

There would have been objections to siting a facility even if it were only for the needs of North Carolina, but the idea that the state was to become a "dumping ground" for seven other states resulted in even greater opposition. The Nuclear Regulatory Commission had issued criteria for site selection and methods of operation that were practical and would result in no more exposure to the public than is received from natural sources during a few minutes each year. The environmentalists wanted even less exposure to the public, and the legislature responded by establishing what most professionals regarded as unnecessarily strict and overly expensive design criteria and protective features beyond those required by the federal government.

North Carolina had a well-organized coalition of environmental organizations that sent representatives to our monthly meetings. Like many states and the federal government, extremely liberal open-meeting laws were in effect that prevented the authority from holding executive sessions unless it was discussing personnel appointments, land purchases, or contractor selection. All other matters were required to be discussed in open session, and the presence of representatives of the media and hostile members of the public greatly inhibited the work of the authority. It soon became apparent that its decisions would be influenced less by the technical needs of the state than by the demands of a few persistent representatives of citizen action organizations that were less interested in helping us find a satisfactory site than in blocking anything having to do with nuclear energy.

After six months, by which time the course the authority should take was reasonably well charted, I began to find the position overly burdensome and resigned. It was taking far more of my time than anyone had anticipated, and I did not enjoy the adversarial relationships that were developing between myself and certain of the environmental activists. I was appointed to the position because of my technical knowledge and experience, but what was required was someone with more political acumen than I ever had had reason to develop.

However, I did not fully disengage from the subject of low-level radioactive waste management. The American Medical Association decided to prepare a statement on the matter because of its concern about the difficulties hospitals and clinics were having in disposing of their waste. I was asked to serve as an advisor to them, and assisted them in preparing the report. I also wrote a short paper on the subject for the *Journal of the North Carolina Academy of Science,* and a somewhat more detailed paper for the New York Academy of Medicine. In all of these articles, my main theme was that existing knowledge makes it possible to design, build, and operate a low-level repository without danger to the public health.[5]

The Galileo Space Mission

In 1987 I was asked to serve on a panel created by the National Aeronautics and Space Administration (NASA) to examine the radiological safety of *Galileo,* a deep space probe that was scheduled to begin a voyage to Jupiter in the fall of 1989.

For about thirty years, both the U.S. and the USSR have been using the energy of radioactive decay to provide thermoelectric power for the electronic equipment aboard spacecraft. I was first associated with the U.S. program in the mid-1950's when the New York Operations Office of the AEC was given responsibility for procuring some of the first such units. However, when I joined the panel, I had had no contact with that program since that time.

Most spacecraft are equipped with solar energy panels from which the power required to operate electronic equipment is supplied. However, these may not be practical for long missions in which the panels are subject to damage by meteorite impacts, or for deep space probes that do not receive sufficient solar energy because of the distance from the sun. For both reasons, it was decided by NASA that a series of contemplated unmanned space missions to the outer planets would be equipped with radioactive thermal generators (RTGs) that would contain about 300,000 curies of plutonium 238, a nuclide with a half-life of 88 years.

The main safety concern was that plutonium might be released to the environment either as a result of a catastrophic launch-pad accident, or because of unintended reentry and burnup in the upper atmosphere. Extensive design precautions were taken to assure that the capsule that contains the plutonium would remain intact in either case. The design calculations were verified in laboratory simulations. As a result, the panel was satisfied that the launch could proceed, because the risk of significant exposure of the public would be exceedingly small.

Despite the panel conclusions, I became concerned that because plutonium decays by emission of alpha particles, it would be difficult to detect in small amounts, and in the event of a mishap it might not be possible to provide the public with data about the levels of exposure in a timely manner. Moreover, while it would be possible to keep the dose to the public well within the limits established by regulatory authorities, there could be no assurance that this would be acceptable to the public. We were in a period in which the public was showing increasing evidence of its unwillingness to accept any exposure to radioactive substances, even though the amounts were insignificant compared to the levels we routinely receive from natural sources. The plutonium power generators were to be launched from Cape Kennedy, where even small levels of general contamination, however insignificant to the public health, could have serious emotional consequences on the residents of Florida. This was another example of my great concern over the disparity that had developed between actual environmental risks and the risks perceived by the public.

TEN

From the Past to the Present

A Fifty-Four Year Perspective

Many careers are influenced by historic events. "There is a tide in the affairs of men, which, taken at the flood, leads on to fortune." In my case I was carried along by three flood tides that took place in mid-century, each of which greatly expanded the opportunities available to me. These were the surge in industrial production with the onset of World War II, the postwar developments in nuclear energy, and the environmental populism that began to develop in the mid-1960s. The years I spent in the war industries and in the postwar atomic energy program have been discussed already; it is also important to understand the history of the modern environmental movement and the great impact it had on my profession and on me.

The popular environmental movement that began in the 1960s has resulted in nothing less than a social revolution. Basic changes in attitudes about the environment have occurred, and the public has become aware that humankind is part of a delicately balanced ecosystem on which all living things depend. There is increased recognition of environmental limitations, and environmental protection has become an accepted obligation of federal and local government. There are new laws, and new agencies for their implementation. New scientific and engineering disciplines have evolved.

The scientists and engineers involved with environmental protection can be divided into two groups, depending on whether they are concerned primarily with the natural environment or human health. The environmental health specialists, for example, are concerned with the effects of radiation from nuclear power plants, or air and water pollution from the chemical industry. The scientists concerned with natural systems might, for example, address preservation of species, impacts of deforestation, or erosion of soil.

There are, of course, significant overlaps, as with contamination of the atmosphere by the oxides of nitrogen and sulfur. These pollutants have implications for both human health and the natural environment since they can increase the prevalence of human respiratory disease and also cause damage to plant life and fish. Similarly, pesticides that persist in the environment can be injurious to both people and wildlife.

My own interests have been confined almost entirely to the harmful effects of technology on human health, and it should be recognized that the discussion that follows is written from that perspective.

ENVIRONMENTAL POPULISM

The present wave of popular interest in environmental health began in the mid-1950s. Until then, the interest in the relationships between the environment and human health were limited mainly to the need for improved hygiene in the traditional sense. Public health specialists were involved with the development and enforcement of regulations concerned with water supply, sewage disposal, housing, and control of communicable diseases. The one exception was in industry, where accidents and occupational diseases began to attract attention earlier in this century.

Concern about the effects of industrial chemicals on human health began to extend beyond the workroom to the general community as a result of a number of air pollution episodes associated with periods of meteorological stagnation. The first of these occurred in 1930 in the Meuse Valley of Belgium. Later, in 1948, an acute air pollution episode in Donora, Pennsylvania was responsible for the deaths of twenty people, and hundreds more became ill. Far more frightening was a 1952 four-day period in London when much of England was covered with a foggy mass of stagnant air. The accumulation of sulfur oxides and smoke caused an estimated thirty-five hundred deaths from acute respiratory disease. The toxic effects of industrial poisons were putting the general public at risk. In my experience, this was also demonstrated by the finding that beryllium poisoning was occurring among residents in the vicinity of an Ohio metal smelter.

Coincidentally, sustained testing of nuclear weapons in the atmosphere began in the early 1950s and demonstrated for the first time that mankind had achieved the ability to contaminate the environment on a global scale. Each of the various radioactive elements contained in fallout moved through the environment in its own way, depending on its chemical and metabolic characteristics. As an example, radioactive iodine and strontium were deposited on grass, from which they were absorbed by foraging cows and secreted in milk. When the contaminated milk was consumed, especially by children, the radioactive iodine was deposited in the thyroid, and the radioactive strontium, because of its chemical similarity to calcium, was deposited in the skeleton.

The bombs being tested in the mid-1950s were already a thousand times more powerful than the bombs dropped on Japan during World War II, and there was speculation that they could be made even more powerful. Krushchev announced in the fall of 1961 that the Soviets were capable of detonating a bomb with an explosive yield of 100 million tons of TNT (nearly ten thousand times the yield of the Hiroshima bomb); the Soviets actually tested a bomb with an explosive yield greater than fifty million tons of TNT. Laboratories all over the world measured traces of the bomb debris in the atmosphere and found some of the radionuclides in food and in the bodies of children. The subject of testing became an emotional subject worldwide and whether to test became a major political issue.

Opposition to nuclear weapons testing was begun originally by those who wanted to stop their development. They were soon joined by others who raised concerns about the dangers of exposing the public to the radioactive materials in fallout. Until the late 1950s it was generally accepted that a threshold radiation dose existed, below which there would be no effects from exposure. However, new laboratory and epidemiological research suggested that there might not be a threshold and that it would be prudent to assume that the risks of producing cancer or causing genetic effects are increased by all radiation exposure, however small the dose. It was further suggested that the dose-response relationship is linear, i.e. a strict proportionality exists between the dose received and the health effects produced. Thus, for every increase in radiation exposure, there is a corresponding increase in risk. At very low doses, of the order of the natural radiation background, the increased risk is exceedingly small but finite nevertheless. This being the case, there is no "safe" dose of radiation. The risks associated with radiation exposure could no longer be based on the answer to the question What is the safe dose? If there is no threshold, there is no such thing as a safe dose. This made it necessary to pose a new question, How safe is safe enough?, a question whose answer has eluded consensus to the present time. Moreover, the same dilemma now exists concerning the effects of the carcinogenic chemicals.

Disagreements about the effects of the fallout developed among scientists, and the subject even became a major issue in the 1956 presidential campaign. Scientists on both sides of the debate had access to the same data and their scientific conclusions were in essential agreement. The basic disagreement was not over the scientific facts. What differed was in the way they presented the data. The two principal protagonists were men of great scientific stature. Linus Pauling was a Nobel Prize chemist who also would later also receive the Nobel Peace Prize for his work in the anti-testing movement. Willard Libby, a member of the Atomic Energy Commission, was a distinguished chemist who would soon receive the Nobel Prize for developing the use of carbon 14 in determining the age of archaeological specimens. Using the best data available at that time, Libby calculated that the risks to the individuals were very small, on the order of one in a million, a level of risk that would not cause many people to be concerned. Pauling multiplied the probability of harm to an individual by the large numbers of people exposed and predicted that a substantial number of cancers would develop. For the world population of four billion people, a risk of one in a million of developing leukemia would result in four thousand cases, surely something about which governments should be concerned.

In August 1963 a treaty was signed that banned testing nuclear weapons in the atmosphere. By coincidence, it was also the year when the first order for a privately owned nuclear power plant was placed by a United States utility. The antinuclear movement, formed to protest the testing and manufacturing of weapons, never lost the momentum it had developed. Some of its most active members began to focus on nuclear power.

The Atomic Energy Commission was in an unfortunate position during the fallout debate. The testing programs it conducted were mandated by government policy, as developed by the president, the Congress, and the State and Defense Departments. The role of the AEC was to implement the policies developed by others, but the onus of testing fell totally on AEC so far as the public was concerned, which caused the agency to lose the trust of the public. The AEC eventually went out of existence and its functions were assigned to two new agencies. The Nuclear Regulatory

Commission (NRC, not to be confused with the National Research Council) was given responsibility for regulatory matters, and the Department of Energy was assigned the AEC's responsibilities for civilian and military research, development, and production. Unhappily, the new agencies have fared no better than the AEC in winning the confidence of the general public.

The year 1963 was destined to be of even greater importance in the history of the environmental movement. In addition to being the year in which the limited nuclear test ban agreement was signed and the first order was placed for a privately owned nuclear power reactor, it was also the year in which Rachel Carson published *Silent Spring*. The book explained the subtle ways in which organic pesticides move through the environment and endanger many forms of life. It was the Carson thesis, presented in eloquent prose, that there could someday be a "silent spring" because the pesticides were poisoning the food being eaten by the birds. The book had an explosive effect on public interest in the environment.

The antinuclear activists were surprised to find that many of their concerns about radioactive fallout were applicable to chemicals as well. Like radioactive fallout, the presence of man-made chemicals could be demonstrated on a global scale. Like the radionuclides, many of the chemicals were also possible human carcinogens that could pass insidiously along ecological pathways.

Global contamination by radioactive fallout could be studied readily because instruments were available that permitted the detection of exceedingly small amounts of radioactivity. Until about 1960, however, this was not so for toxic chemicals because the methods of detection were tedious and expensive. Development of the atomic absorption spectrophotometer suddenly simplified the analyses for trace elements such as lead, arsenic, and mercury. At approximately the same time, improvements in the techniques of gas chromatography and mass spectrometry increased the ease of detection of traces of complicated organic molecules, such as DDT and the family of polychlorinated biphenyls (PCBs). With these convenient methods of analysis, it was soon found that not only were many toxic substances distributed more widely in the environment than had formerly been realized but that some of them were being concentrated manyfold as they ascended the ecological food chains. Seemingly insignificant amounts of DDT in pondwater were found to increase step by step, from phytoplankton, to zooplankton, crustaceans, small fish, larger fish, and so on to human beings and the other mammals and birds at the top of the food chain. The most alarming finding was that the DDT was concentrating in the tissues of birds to such an extent that their ability to reproduce was being impaired.

Silent Spring was unquestionably the single most influential factor in the sudden development of environmental populism. The public became aroused, environmental societies were formed in hundreds of communities, and some of the larger organizations developed powerful voices in the media as well as in legislative and judicial chambers. The widespread public concern resulted in political action and in the decade that followed the legislative basis for much-needed programs of environmental protection was established at the state and federal levels. Huge amounts of money were appropriated for research, and the Federal Clean Air and Clean Water Acts were passed. Within a few years a revolution had taken place in society's understanding of the environment and the need to protect it. There developed a general awareness that the physical and biological bits and pieces of the world are linked together in delicately balanced ecological networks, and that damage to one component of the ecosystem might affect the system as a whole.

The environmental movement was a true grass-roots development not anticipated by our national leaders. In preparation for the administration that would be formed after the election of 1968, the prestigious Brookings Institution assembled a panel of twenty of the country's leaders to prepare a book, *Agenda for the Nation,* that discussed the subjects that the authors believed would be important during the coming years. The ex-cabinet officers, agency heads, and academic leaders selected eighteen subjects for discussion, including housing, military preparedness, and poverty, among others. The environment was not included, presumably because it was not considered sufficiently important. Yet Richard Nixon's first term (1968–72) proved to be a period of intense environmental activism, as a result of which the National Environmental Policy Act specifying the requirements for environmental impact analysis was signed into law, and many new agencies were created, including the Environmental Protection Agency and the Occupational Health and Safety Administration.

The beginning of the environmental movement was exciting to the environmental health specialists of my generation. People other than ourselves were sharing our concerns for the first time. The words "ecology" and "environment" entered the popular vocabulary, and there was much attention given to these subjects in the electronic and printed media. The federal government was given major responsibility for environmental protection, and large sums of money were made available for air and water pollution control. In the twenty-five years that preceded the onset of the environmental movement I never dreamed that the field in which I was working would attract so much attention. The world developed an environmental consciousness and began to appreciate that the human species is part of an ecological system within which each component is linked to every other. People realized that each generation is the temporary steward of the environment and has the responsibility to preserve it for the future. The environmental awareness that has developed, and the ways in which the government and industry have restructured themselves to deal with environmental problems, have constituted a historic social change.

Many of the new environmentalists, as they began to be called, were activists with considerable charisma. As a group they tended to be apocalyptic in their writings. Books were published with titles such as *Famine, 1975*, and *Vanishing Air*. Wave after wave of excitement about environmental matters spread through the country. In quick succession, the public soon learned about mercury in fish, carbon monoxide, a long list of carcinogens in foods, and black lung disease among Appalachian miners. At the beginning of the environmental movement, the subject was so popular that the newspapers and magazines regularly included special sections on the subject. Some environmental organizations focussed on roadside litter and nonreturnable bottles, and others on the need to save the whales or the desirability of zero energy growth. It has been estimated that by 1973 there were five thousand environmental organizations in the United States.

ENVIRONMENTAL PRIORITIES

I soon began to share with many of my professional associates a concern that many of the environmental populists were failing to comprehend the complexities of the issues with which they were dealing. I concluded that priorities were becoming disordered. Furthermore, there was at first a persistent antitechnology tone to the environmental movement. A return to the simple life of the past began to be a dominant theme, particularly among the younger members of the environmental movement. Small numbers of them lived in communes where they practiced organic farming and used no electricity.

Moreover, most of the environmental issues that were attracting attention at the height of the environmental excitement were superficial and not threatening to society. Despite all of the environmental problems that were developing, most people were living in better homes, were safer and more comfortable at work, were better nourished with wholesome food, and were even safer on the highways than when I first became concerned with such matters. More people were also drinking wholesome water and breathing clean air. In many cities air pollution control programs had already been underway for many years. In particular, there had been substantial reductions in the amounts of soot and sulfur oxides in the air. Blue skies had become the rule rather than the exception in cities such as Pittsburgh, London, and St. Louis that had long been known for their dense smog. Although much remained to be done, progress had been made.

The subject of the environment was becoming ill-defined and reports that appeared daily in the media confused the public. Suddenly there seemed to be a new crisis each week. These often dealt with problems of which professional environmental health specialists were well aware, but the media described them in an exaggerated and alarming tone. Some scientists who had not been previously involved with such matters were proclaiming that Lake Erie was dead, that we should generate electrical energy from solar energy and wind power, and that we should recycle most of our waste. Most such statements contained an element of truth: Lake Erie was badly polluted, but was far from "dead"; it was an admirable objective to obtain electric energy from renewable sources, but this was not then or now practical on a large scale; and it was desirable to recycle wastes, but only when it was practical to do so. A basic problem was that the new environmental spokesmen wanted simple and quick answers to complicated questions, which simply was not possible.

The fervor of the environmental movement at the height of public interest was symptomatic of the times—a turbulent period and an era of popular causes. There was violent opposition to the war in Vietnam. Cities were burning as a result of racial tension. The women's liberation movement was getting underway, and fundamental changes in sexual mores were taking place. The popular movements of the sixties and seventies stirred intense emotions and the environmental movement was no exception.

In my concern about the superficiality of popular environmental issues, I believed that the public was neglecting more basic problems. I was particularly concerned with an overemphasis on pollution, particularly that of industrial origin. Pollution control was my specialty so I knew very well the seriousness of the problems that existed and how much needed to be done. However, I considered environmental pollution secondary to far more basic problems.

In 1970, at the height of the new excitement, I was invited to prepare a paper on "Technology and Man" for formal presentation at a small weekend meeting in England that included about thirty leaders in environmental protection from the U.S. and U.K. After reviewing how much technology had contributed to the well-being of mankind in the developed part of the world, I then identified what I thought were the six most important environmental problems. These, in the order of the relative importance I attached to them, were (1) nuclear war, (2) exponential growth of population, (3) waste of natural resources, (4) inadequacies in urban design and organization, (5) poverty, and (6) pollution.[1]

Pollution was last on my list, at the top of which I place nuclear war. This must have seemed strange to some, considering that my life had been spent in efforts to control pollution. However, the threat of nuclear war was and is unquestionably the overriding threat to the human environment. Population growth comes next because it is creating the other problems, including pollution. The human environment will be seriously affected if we deplete critically needed natural resources. Nor can society function if the urban infrastructures are allowed to deteriorate, as seems to be happening in many cities. The environmental concerns of people who live in poverty have little in common with those who are more fortunate and can place high priorities on clean air, clean water, and litter-free streets. Rats, roaches, and dilapidated housing are very much a part of the human environment. Pollution, I concluded, is the only one of the six for whose management we have both the knowledge and the social institutions.

During the two years I spent with Mayor Lindsay in New York City, I had developed a deep appreciation of the environmental problems of the urban poor. When considering their needs, the environmental movement seemed elitist. The activists on the college campuses knew little about the environmental problems of poor people. I was concerned that the environmental movement was focusing on the things that seemed important to the privileged people in society, and that the environment in which the urban poor lived was being neglected.

It is now nearly twenty years since I prepared that paper for presentation in England. I am still convinced that the dangers of pollution are greatly exaggerated in the public mind. This is particularly true so far as health effects are concerned. There are examples of injuries and death caused by pollution, but they have been comparatively rare and should be even more rare in the future because of precautions that have been adopted. The greatest danger from pollution is the threatened increase in global temperature resulting from the increasing carbon dioxide content of the atmosphere. This possibility has not, either then or now, alarmed the public nearly so much as the presence of trace contaminants in air, food, and water.

Environmental fervor reached a climax in April 1970 with nationwide celebrations of Earth Day, which was devoted to teach-ins in schools, churches, and public meeting places. I was invited to address convocations on several campuses. Two of my addresses followed addresses by Barry Commoner, a plant pathologist turned ecologist-philosopher-politician. (He was later a candidate for president of the United States on the ticket of a party he founded.) Commoner deplored the technological society in which the students lived and reminded them that they were the first generation in history to have PCBs and DDT in their bodies. In contrast, I called their attention to the benefits they were receiving because they lived in the modern world. They were enjoying privileges that were not available even to kings in former times. The students were, I emphasized, part of the first fortunate generation of people whose blood contained the antibodies that protect against diseases such as poliomyelitis and diphtheria, diseases that were scourges in the times of their parents and grandparents. It was a fair trade, I thought, to be carrying traces of contaminants that were doing no measurable harm in exchange for the many benefits provided by modern technology.

The fears about the health effects of pollution are resulting in the waste of huge amounts of money. I have previously discussed two such examples—the design criteria for the North River Water Pollution Control Plant and the air quality standard for sulfur dioxide in New York City. Those two decisions involved expenditures of hundreds of millions of dollars above what was required to protect public health. The money could have been better spent on traditional problems of municipal sanitation such as vermin control, or on school lunches or health education. Nationwide, steps that were costing many billions of dollars were being taken to control pollutants that were having only marginal effects on health at the expense of other public health problems that were badly in need of correction.

ENVIRONMENTAL CANCER

Nowhere in the field of environmental health is the public more confused than about the relationship between the environment and cancer. For twenty years, there have been repeated reports that various substances are suspected of causing human cancer. Many of the indicted products have been well known to the public, such as coffee, artificial sweeteners, various food additives and pharmaceuticals. There are frequent reports that traces of carcinogens have been detected in food products.

Cancer is a dreaded disease that causes one out of four deaths in the United States. In a 1964 report, the World Health Organization stated that as many as 60 to 90 percent of all cancers might be the result of some environmental factor. That report was issued only one year after publication of Rachel Carson's book, *Silent Spring,* which warned about the dangers of polluting the environment with toxic industrial chemicals, some of which were known to produce cancer in experimental animals. Having been sensitized to the issue by Rachel Carson, some people erroneously interpreted the WHO report to mean that 60 to 90 percent of all cancers were the result of industrial pollution of air, water, and food. This was not intended by the writers of the report, which referred to the environment in the far more general sense of the circumstances in which a person lives. Cigarette smoking was the most important environmental factor identified, but personal hygiene, methods of preparing food, occupational exposure to carcinogens and even sexual practices were also cited as important risk factors. Environmental pollution was properly included as a possible factor, but there was no intention of implying that it was the major cause of cancer. Nevertheless, the WHO report was widely cited to promote fear of chemicals.

It is true that the number of deaths from cancer has been increasing from year to year, but an important reason for this has been the increasing size and average age of the population. Many people who in former years would have died at a young age from communicable diseases are now living to old age. This increases the probability that they will eventually be afflicted by cancer, the incidence of which increases rapidly with age. To correct for these influences it is necessary to normalize the statistics, which is done annually by the American Cancer Society. The cancer trends illustrated in figures 2a and 2b show that during the past fifty years the only significant increase has been for lung cancer in both men and women. This is due almost entirely to the practice of smoking cigarettes, which increased in popularity among men during the World War I era. Because of the delay in the development of this disease, lung cancer did not begin to rise rapidly among males until the mid 1930s. In 1989 the disease was estimated by the American Cancer Society to account for 35 percent of the 266,000 deaths from all cancers among men in the United States.

The popularity of smoking among women lagged behind that of men by about the period between the two world wars. As shown in figure 2b, lung cancer began to increase among women in the post-World War II period. Since 1985 the death rate from this self-induced disease has exceeded that from breast cancer, the most dreaded of all diseases of women.

It is seen that for both males and females there have been no dramatic increases in cancer mortality except for lung cancer. For reasons that are not fully understood, there has been a remarkable reduction in cancer of the uterus among women and in stomach cancer both in men and women.

Legislators and government agencies had to respond to the concerns of the public which, by 1970, wanted clean water and clean air, regardless of price, because of the belief that pollution was the major cause of cancer. The public wanted no detectable chemicals in food, however low the concentration, if the chemical was known to produce cancer in any species of experimental animal, at any level of exposure, however great. (This prohibition had already been incorporated into law by the Delaney Amendment to the Food, Drug, and Cosmetic Act in 1958.)

By 1978 I believed that exaggerated fears about carcinogens in the environment were causing the priorities for environmental protection to become distorted. In response to this concern, I published an article on environmental cancer in the magazine *Environment*.[2] It also was a chapter in my book, *Environment, Technology, and Health,* which was a broad review of the field that placed the subject of human ecology in historical perspective.[3] I have been pleased that my conclusions conformed to those of more recent publications by epidemiologists and cancer specialists who have been much closer to the subject than I.

The great gap that exists between the actual and perceived dangers from exposure to traces of carcinogenic substances has continued to widen until the present time. Nowhere is it so wide as with concerns about exposure to traces of radioactivity, which has resulted in enormous disparity in the costs society is willing to pay to avert cancer. Thus it was estimated in 1980 that, on average, one death from cancer of the cervix could be prevented by spending $25,000 for education and screening, and that accidental death from smoke inhalation could be prevented at an average societal cost of $40,000 by requiring that smoke detectors be installed in bedrooms. Many other examples could be given of ways in which premature death can be prevented at modest cost.

However, far greater costs are incurred for premature death prevention associated with the effects of modern technology. The estimates range to hundreds of millions of dollars per prevented death from exposure to chemical carcinogens, and to billions of dollars to avert one premature death from radiation exposure.[4] The disparities are illustrated by the following example.

At Three Mile Island in Pennsylvania, 2.3 million gallons of water contaminated with slight traces of radioactivity have been accumulated in the course of cleanup activities. The water contains tritium and traces of other radionuclides in such small amounts that it could be discharged to the Susquehanna River without exceeding the limits prescribed by federal regulations. However, this has not been done because of the opposition that has developed in nearby communities. The National Council on Radiation Protection and Measurements evaluated the potential health effects of discharging the water into the river and concluded that the dose to the maximally exposed person would be 2 microrem. This is equivalent to the dose received from about four minutes of exposure to the radiation present naturally in the environment due to cosmic rays and radioactive materials in the earth's crust. The collective dose to the population (the mean dose times the number of people exposed) was calculated to be about 1 person-rem. Because of local opposition, it was decided that the water should be evaporated at a cost of about five million dollars. Since the risk of developing cancer from radiation exposure is estimated to be, at a maximum, about 2 per 10,000 person-rem, the cost of averting one case of fatal cancer is about $25 billion dollars.[5]

No doubt there are people who find it repugnant that actions taken to

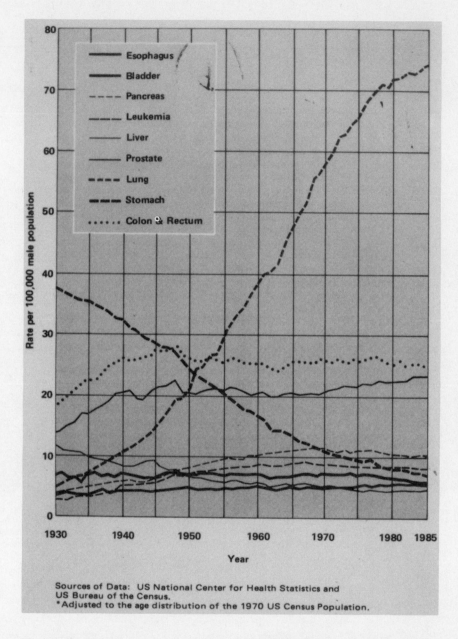

Fig. 2a. Age-adjusted cancer death rates for males in the United States for selected sites, 1930–1985. Death-rate figures are adjusted to the age distribution of the 1970 U.S. Census Population. Source of data: U.S. National Center for Health Statistics, U.S. Bureau of the Census, and the American Cancer Society.

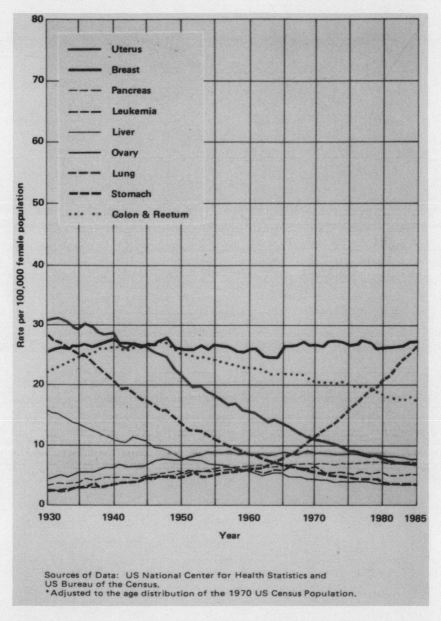

Fig. 2b. Age-adjusted cancer death rates for females in the United States for selected sites, 1930–1985. Death-rate figures are adjusted to the age distribution of the 1970 U.S. Census Population. Source of data: U.S. National Center for Health Statistics, U.S. Bureau of the Census, and the American Cancer Society.

prevent premature death should be based on the cost of doing so, but there is no alternative in a society where there are limited resources and so much to do. Hiring an additional school nurse, construction of a new firehouse, or implementation of an educational program to assure immunization of children all require funds that must be provided in competition with other needs.

The Media and Excessive Public Concern

This book is not the place for a detailed analysis of why the public is excessively concerned about environmental carcinogens (and other pollutants as well), but a few observations are in order. We must start with the fact that a small minority of the population is unwilling to agree that people should be exposed to any risk, however small, or however large the benefits. These are people with mind-sets of the kind that have blocked water fluoridation for nearly fifty years in many communities despite the great benefits that have been demonstrated with no demonstrable human effects. They are a minority of people that do not understand the tradeoffs between risks and benefits and want one without the other.

The publications and public statements made by the activists and their organizations are often dramatic and are regularly publicized by the media. These reports are far more apt to be publicized than those of a more assuring tone published in the scientific journals. The media treats news sensationally because that is what the public wants it to do. The number of newspapers and magazines sold and the numbers of TV program viewers are the only indices of media success. There are no indices of accuracy of presentation.

Because of the tendency of the media accounts to be sensational, scientists of moderate views are rarely interviewed for their opinions or are likely to refuse interviews if requested. The latter is an unfortunate source of bias that results from the poor quality of the reports based on interviews the scientists have given in the past. Other reasons why most scientists develop an aversion to the press are the following:

1. Most reporters don't prepare properly for assignments on technical matters and often do not know enough about the subject to ask the right questions. They call on short notice, with deadlines that are frequently only three or four hours away. (This is generally not true of

the science reporters, but in recent years they have covered very few of the stories about environmental health.)

2. Many knowledgeable scientists will not grant interviews because they know from experience that they are likely to be quoted out of context. This is particularly true for television where the scientist is not consulted about the portions of the interview that are selected for showing. All too frequently the best part of the interview is edited out of the tape.

3. It has become commonplace for the media to give publicity to individuals who believe they have been injured by some toxic chemical or radiation. Examples have been people who lived near the Love Canal in New York, residents from the vicinity of the Three Mile Island accident, and neighbors of numerous industrial and research facilities. The tragedies in the lives of these unfortunate people are exploited by the media and abetted by activists, despite epidemiological studies that fail to demonstrate causal relationships between the complaints and the alleged source of exposure.

The media influences what the public thinks which, in turn, influences the actions taken by government. Elected officials are expected to reflect what the public wants, however ill informed the public may be. This in turn has an influence on decisions made by the agencies charged with responsibility for environmental matters. The agencies are expected to adopt policies that are consistent with the wishes of both the public and the legislators who represent them. An agency must avoid taking issue with the legislators, because to do so would jeopardize the financial support that only the legislators can provide. For this reason, a scientist who disagrees with a proposed agency position is not likely to be appointed to either its staff or advisory committees. The scientists cannot be picked on the basis of their competence, but primarily on whether they support the policies the agency wants to develop. This explains why the scientists become polarized. The "ins" continue to support the policies of the agency, whereas the "outs," who may be capable, are likely to be sought as industry consultants. In the process, the agency loses its independence.

I have been disappointed that so many members of the scientific community have remained silent for so many years when they could have helped the public place the subject of environmental cancer into proper perspective. Few scientists consider it to be their role to improve public understanding of the subject and the vast majority of scientists have allowed their few colleagues with extreme views to dominate communications with the media and the public.

FAILURE TO ADDRESS BASIC PROBLEMS

My greatest disappointment has been that the nation's policies have not addressed the most basic of the environmental problems. The environment presents myriad aspects and is perceived by many people from their own vantage point. In many ways the environmental policies of the nation have been designed to deal with secondary problems that may be the result of more basic problems. Chemical dumps, asbestos, automobile engine emissions, and nuclear wastes are examples of matters that clearly require attention, but they have dominated the public attention to such an extent as to detract attention from more fundamental problems. Examples of basic problems are the need to control population size, to assure a supply of energy, and to prevent atmospheric pollution that may alter the global climate.

Population Growth and Environmental Problems

The problems created by population growth can be appreciated readily by considering that the time required to double the world's population has become progressively shorter over the centuries. The global population is estimated to have consisted of only 250 million people at the start of the Christian era, and it took until about 1650 to double in size and reach 500 million. It then took only 150 years to double again and reached one billion by 1800, and then reached two billion people in 1900. As we approach the end of the twentieth century the world's population stands at about five billion, with a projected doubling time of about thirty-five years. This kind of exponential growth will soon have disastrous consequences. Most of the growth is taking place in the third world where at this writing the doubling time for its 2.8 billion people is currently estimated to be about thirty years.

The situation is far less urgent in the developed countries, where near stability has been achieved. The populations of some of the countries of Western Europe are actually diminishing. In the United States, where the population was only 76 million at the start of the twentieth century, it is currently about 250 million and growing at the rate of about 0.7 percent per year. At this rate, the country's population could approach 500 million by the end of the twenty-first century. This would pose enormous environmental problems. The pressure from our burgeoning population can no longer be relieved, as early in this century, by moving west. When we consider the problems we endure because of congested roads, our inability to dispose properly of wastes, and our chronic inability to find suitable sites for housing and industrial facilities, the prospect of having to accommodate twice the present population is frightening.

An additional complication is that the rate of population growth is not occurring at an even rate throughout the U.S. population. Birth rates are highest among the disadvantaged people who have the lowest level of education. Moreover, the growth in our indigenous population is augmented by an annual influx of perhaps one million or more immigrants from the less developed countries, particularly Latin America. This is a highly significant development that can be interpreted as being due to the failure of those countries to achieve a balance between the natural resources they possess and the number of people they must support. In the past we could welcome such people with open arms because we were a growing nation that needed people to build the roads and railroads and work in the mines and mills. That is no longer the case. What does a lifeboat captain do who knows there are still people to be rescued, but realizes that his boat is so full of people that it will be swamped if others are taken aboard?

It is in our national interest that there be a population policy based on the fact that the quality of the environment is clearly related to the number of people that it must support. To the contrary, our government has placed its head in the sand. Funds for sex education have been curtailed and the government in recent years has strongly opposed abortions. There has been no recognition of the fact that population pressure is bound to degrade the quality of life for our citizens of the twenty-first century. Unrestrained population growth must be regarded as the primary threat to the environment.

The importance of population stabilization in assuring both the quality of life and the health of the natural environment is understood by many environmentalists but their efforts have not generated the zeal required to influence national and international policies. Activists go to Brazil to help preserve the trees of the Amazon, not to help slow that country's birth rate. Many such examples could be given.

The explosive growth of population as well as the related problem of increasing affluence and material expectations have been in part the result of technological developments during the past century. The phenomenon serves as an excellent example of the tradeoffs that occur between benefits and detriments when new technology is introduced. Technology has made it possible for people to live better, with longer lifespans, less sickness, greater wealth, and more leisure. However, technology in the less developed countries has substantially reduced death rates without compensating reductions in births. As a consequence, many of the developing countries are in the midst of a population explosion that is having disastrous effects on both human well-being and the natural environment. In the more developed countries, where the populations are generally stabilized, technology has increased human affluence to such an extent that people are consuming more goods, using more energy, and creating more wastes. Thus, both the population increases of the less-developed countries and the rising demands for goods and services in the developed countries are producing more pollution, diminishing natural resources, and placing increased stress on the natural environment. Both are examples of the difficulty of foreseeing the eventual environmental impact of the most benign technological innovations.

Electrical Power and the Quality of Life

Human productivity during the past century has increased sevenfold in the U.S. despite the fact that the work week has been reduced from about sixty hours to forty. People are better fed, better housed, and have more wealth with which to enjoy more hours of leisure. People live longer and, perhaps more important, suffer fewer days of disability during their productive years. A person afflicted with tuberculosis early in this century not only was likely to die at a young age, but also would probably suffer many long weeks of incapacity.

The great improvements that have taken place in the quality of life in the developed nations of the world could not have occurred were it not for the ready availability of inexpensive, clean, and convenient electric power. It is electrical power that has made it possible to create the great variety of goods and services on which we depend for a productive and comfortable life. Electricity has reduced the manual effort required both in the home and at work. It is required for communication, illumination, computing, and sophisticated manufacturing procedures. Even the miracle drugs that have eliminated so many of the diseases of former times require electricity for their production. Without electric power it would not be possible to manufacture those pharmaceuticals in sufficient quantity and to the required high standards of chemical and biological quality. Nor could we even grow the food we need. The high productivity of modern agriculture requires electrical energy to produce synthetic fertilizers, pesticides, and farm machinery. Energy is also required to process, distribute, package, and store the food produced. In short, electric power is essential for the operation of a modern economy.

Among the nations of the third world there is correlation between per capita energy use and well-being as measured by life expectancy at birth. Among all the nations, per capita energy use is closely correlated with per capita gross national product. Poor people in the have-not countries use less energy and on average do not live as long as people in the developed countries.

Since energy use is so intimately associated with well-being, it is indefensible that our country has not had a national energy policy. Such a policy is needed to assure that cheap energy will continue to be available in sufficient quantity, and that it is produced and used in an environmentally acceptable manner. The U.S. will need more energy in the future, not only to maintain the present standard of living of its more fortunate citizens, but also to allow improvement in the quality of life for those who are now disadvantaged.

Environmental Impacts of the Fossil Fuels

Coal, oil, and the other fossil fuels are our main sources of energy. There are abundant supplies of coal but its mining results in spoiled landscapes and injuries to workers. When coal is burned, the combustion products that pollute the atmosphere are injurious to human health and can be damaging to plant life. Petroleum combustion is less damaging to the landscape and has fewer effects on human health but transportation of the liquid fuels by tankers has been a continuing source of concern because of oil spills that have damaging ecological effects. In addition, offshore drilling also involves the possibility of oil spills and is a dangerous occupation, particularly because the drilling platforms on which the men live and work are subject to destruction by storms and heavy seas. Another problem with petroleum is the uncertainty of supply because of political instability of the oil-rich nations of the Persian Gulf.

The most ominous of the environmental impacts of fossil fuel combustion is unquestionably the "greenhouse effect." It has been known for a century that combustion of fossil fuels is increasing the carbon dioxide content of the atmosphere. This gas is a poor transmitter of the infrared radiation produced when visible light is absorbed at the earth's surface. The presence of carbon dioxide in the atmosphere thus slows the rate at which solar energy absorbed by earth can be re-radiated back into space. The glass in a greenhouse has the same effect. The visible light passes easily through glass, but not the infrared (heat). The greenhouse is warmed because the energy contained in the visible portion of the sunlight spectrum cannot leave it after it has degraded to infrared.

The carbon dioxide content of the atmosphere has increased by more than twenty percent since the start of the industrial revolution and will continue to increase unless fossil fuel combustion is curtailed. The global warming that will result is uncertain but it could be sufficient to result in the melting of enough polar ice to cause a damaging rise in the sea level. The temperature rise might also be enough to cause profound and perhaps disastrous effects on the global climate beginning sometime in the twenty-first century.

The Importance of the Nuclear Power Option

Nuclear energy has been viable as a source of electricity for about twenty-five years, and many studies have demonstrated that it is the most benign of all available energy alternatives with respect to environmental effects. Some countries of the world that want to assure a supply of clean and inexpensive energy are depending on nuclear power as their principle source. By 1987 France derived 70 percent of its electricity from nuclear sources, Belgium 67 percent, and Sweden 45 percent. In the United States, where there has been intense opposition to its development, only 18 percent of the electricity was derived from nuclear power, but in 1987 this was sufficient to replace the equivalent of 200 million tons of coal or 700 million barrels of oil.

There are some people who would have us replace the fossil fuels with solar or wind energy, but these will not be practical in sufficient quantity for the foreseeable future. Nuclear energy is the only viable alternative to the fossil fuels. It would not only eliminate the air pollution from power plants that burn coal, but in time would also eliminate the pollution caused by combustion of petroleum products in automobiles and trucks.

Improvements in the performance of electric batteries should eventually make it possible to replace the internal combustion and diesel engines.

Contrary to popular perception, the nuclear industry has had a good safety record. There are now more than 400 nuclear power plants in the world, and experience with them has provided an ample body of information from which certain conclusions can be drawn.

We know that under normal conditions of operation, the reactors release small quantities of radioactive gases and liquids to the environment, and that solid wastes are produced as well. However, the doses received from these releases by people who live in the vicinity of the plants are well within the limits allowed by regulatory authorities and are minuscule compared to those ordinarily received from nature. The additional radiation exposure to people who live close to reactors is no greater than the increment received by a New Yorker who chooses to live on Manhattan Island rather than across the East River in Queens. (The dose in Manhattan is higher by about 15 mrem per year because of the greater radioactivity of the igneous rock from which most of the island is formed.) People who live in the mountain states may receive twice the exposure of people who live on the coastal plants. I mentioned earlier that the emissions from coal-burning power plants include naturally present radioactive dusts in amounts comparable to the emissions from power reactors.

Some people are of course more concerned about the dangers from accidents than routine operation. Here too the record is assuring. In the forty-five year history of the world's nuclear industry there have been fourteen accidents in which a reactor core has been damaged. This is the kind of accident that has the potential to release radioactive materials to the environment. In only two of these were more than insignificant quantities of radioactivity released. These were from a military (plutonium production) reactor in England in 1957, and from the Soviet nuclear power plant at Chernobyl in 1986. There have been two accidents involving civilian nuclear power plants in the U.S., but neither resulted in significant quantities of radioactive materials being released.

The accident at Chernobyl resulted in the death of thirty-one workers. Although there were no immediate casualties among nearby residents, about 100,000 people who lived within fifty kilometers of Chernobyl received sufficiently high doses that Soviet health authorities have begun studies to determine whether the exposures will result in a higher than normal number of genetic or carcinogenic effects in future years. In addition, clouds of radioactive gases and dust that were released to the atmosphere resulted in fallout that exposed the populations of the USSR and other European countries to doses comparable to those received during the era of open air weapons testing. There may be delayed effects in these populations but at worst, they would be too small to be detectable. Here we are faced with the previously noted dilemma posed when very small individual risk estimates are applied to very large populations. It has been noted that the total projected delayed effects from the Chernobyl accident will be less than the delayed health effects from atmospheric emissions of fossil-fuel-burning power plants in the Soviet Union in a single year.[6]

The accident at Three Mile Island in 1979 destroyed the reactor core, but the protective features designed into the plant prevented all but a small fraction of the radioactivity from escaping to the environment. The highest doses, delivered to a few people who lived near the plant, were less than 100 mrem, which is less than the average dose received from natural sources in about one year.[7] The accident at TMI, in contrast to the one at Chernobyl, demonstrated the importance of the safeguards employed in the U.S. reactor designs. These were largely absent at Chernobyl.

Another frequently stated objection to nuclear power is that there is no satisfactory solution to the problem of how to dispose of waste products. I have mentioned earlier how minute the exposures are from the storage of low-level wastes that originate not only from nuclear power plants, but from hospitals and research laboratories as well. The same is true of the high-level wastes that are the residues from fuel reprocessing, or the fuel itself if it is not reprocessed. Environmental impact statements have been prepared that estimate the risks to the public. For the Waste Isolation Pilot Plant being completed in New Mexico, it is estimated that the lifetime dose to nearby residents would be about one ten-billionth of a rem. This is equivalent to the dose received in a small fraction of one second from natural sources of exposure.

When presented with the facts about the levels of exposure associated with radioactive waste disposal, interveners often say they are unwilling to accept any level of risk, however small. There is no risk-free method of producing energy. It should be sufficient to explain that the risks of low-level radiation exposure, if they exist at all, are so small that they cannot be measured, even with the best of clinical and epidemiological methods. Why should one be concerned about risks that are too small to measure, when there are much greater risks in everyday life that are far more deserving of attention. The answer seems to be that the public will not treat the risks from radiation exposure in the same way as other risks. We have seen earlier that society demands a double standard that will mandate huge costs to further reduce radiation exposures that are already minuscule, while accepting with equanimity much greater risks that could easily be reduced or eliminated.

In former times, the acceptability of a risk being imposed on the public would be determined by the government in consultation with experts. The appropriate agencies of government, the National Research Council, the National Council on Radiation Protection and Measurements, committees of the professional societies, and individual experts, were allowed to decide the course of action that best served the public interest. This is no longer so. The citizens are distrustful, and will no longer delegate to others the power to make decisions that affect their lives. The public has lost confidence in government and in experts. The judgement of leaders was found to be serious flawed in the Vietnam years and when officials lied about Watergate and the Iran-Contra affair. Scientists and engineers failed to prevent accidents to the space shuttle *Challenger,* to the chemical plant in Bhopal, and to the reactors at Three Mile Island and Chernobyl. Do these failures of government and technology mean that those institutions are now less reliable than in the past? I think not. Scandals have rocked government in all times. Catastrophic dam failures, chemical explosions, ship sinkings, mine collapses, and fires were far more prevalent in the first decades of this century than now. The collapse of a dam in Johnstown, Pennsylvania in 1889 drowned 2,200 people. In the last fifty years there have been only two mine accidents in this country that killed more than a hundred men, compared to eighteen such accidents during the previous fifty years. The chronicles of disaster provide many such comparisons. Perhaps the main difference is that in the earlier period the disasters were not televised into our homes.

A Time of Reappraisal

Despite my belief that the environmental health priorities have become misaligned, and that the public has become overly concerned about relatively trivial environmental hazards, the popular movement of the past twenty years has been enormously beneficial. Environmental protection has become an accepted responsibility of both government and the private sector. The principles of environmental impact analysis are being applied, and the legislative framework needed for environmental regulation has been established. Most of all, there is a public awareness of the importance of environmental protection.

In a democracy it is sometimes necessary to achieve new goals by language that is so strident as to be irritating to those who, like myself, approach their objectives with caution. However, the caution inherent in the scientific approach is no doubt even more irritating to the less patient environmental activists.

At the start of this century activists were sometimes called "muckrakers," a word coined by President Theodore Roosevelt. He was greatly influenced by one muckraker, Upton Sinclair, who wrote *The Jungle,* a book that described the appalling practices that existed in the meat packing industry. Aided by a groundswell of popular revulsion as a result of that publication, Roosevelt pressed successfully for federal regulation, and Congress enacted the first food and drug laws. At about the same time Roosevelt said, "The men with muck rakes are often indispensable to the well being of society, but only if they know when to stop raking the muck." Unfortunately, the word "muckraker" has developed a pejorative connotation since it was first used, nearly one century ago. This is regrettable; I also believe that muckraking often does some good.

The raking of muck has produced the sensational and misleading reports about the environment that daily reach the public via the electronic and printed media. This has been disruptive and costly, but it has not been all bad. Good things have been done that could not have been accomplished by more logical means in the illogical world in which we live. However, the environmental activists have by now made their point. Their message has been heard and acted upon. Perhaps Roosevelt was right when he said there is a time to put away the rakes. The time has come for a period of collective introspection to allow us to reorder the environmental health priorities that have become so badly confused.

NOTES

1   RANDOM WALKS AND CHANCE ENCOUNTERS

1. M. Eisenbud, *Environmental Radioactivity*, 3d ed. (Orlando, Florida: Academic Press, 1987).

2. National Research Council *[Report] Committee on Dose Assignment and Reconstruction for Service Personnel at Nuclear Weapons Test*, M. Eisenbud, Chairman (Washington, D.C.: National Academy Press, 1985).

3. B. L. Turner II, ed., *The Earth as Transformed by Human Action*, (New York: Cambridge University Press, In Press).

3   INDUSTRIAL HYGIENE IN THE WAR INDUSTRIES

1. M. Eisenbud, "Patent No. 2,388,604," (U.S. Patent Office, November 6, 1945).

2. M. Eisenbud and L. Silverman, "The Comparative Performance of Air-Supplied Welding Helmets," *Welding Journal*, (1948).

3. C. R. Williams, M. Eisenbud and S. E. Pihl, "Mercury Exposures in Dry Battery Manufacture," *Journal of Industrial Hygiene and Toxicology*, 29(1947): 378–81.

4. M. Eisenbud, "Health Hazards in Aircraft Manufacturing," *Industrial Medicine*, 18(1949): 99–102; M. Eisenbud, "The Principal Health Hazards in Metal Finishing Departments and Their Control," *Metal Finishing*, (1942).

5. R. J. Levine and M. Eisenbud, "Have We Overlooked Important Cohorts for Follow-up Studies? Report of Chemical Industry Institute of Toxicology Conference of World War II Era Industrial Health Specialists," *Journal of Occupational Medicine*, 30(1988): 855–60.

4   AEC: DEVELOPMENT OF THE HEALTH AND SAFETY LABORATORY

1. M. Eisenbud, "Commentary and Update: Chemical Pneumonia in Workers Extracting Beryllium Oxide," *Cleveland Clinic Quarterly*, 51(1984): 441–47.

2. M. Eisenbud, C. F. Berghout and L. T. Steadman, "Environmental Studies

in Plants and Laboratories Using Beryllium: The Acute Disease," *Journal of Industrial Hygiene and Toxicology,* 30(1948): 281–85; M. Eisenbud, "Origins of Standards of Control of Beryllium Disease (1947–1949)," *Environmental Research,* 27(1982): 79–88.

3. R. H. Hall, J. K. Scott, S. Laskin, C. A. Stroud, and H. E. Stockinger, "Acute Toxicity of Inhaled Beryllium," *Archives of Industrial Hygiene and Occupational Medicine,* 2(1950): 25.

4. M. Eisenbud and J. Lisson, "Epidemiological Aspects of Beryllium-Induced Nonmalignant Lung Disease: A 30-Year Update," *Journal of Occupational Medicine,* 25(1983): 196–202.

5. M. Eisenbud, R. C. Wanta, C. Dustan, L. T. Steadman, W. B. Harris and B. S. Wolf, "Non-Occupational Berylliosis," *Journal of Industrial Hygiene and Toxicology,* 31(1949): 282–94.

6. M. Eisenbud, "Health Hazards from Beryllium," *The Metal Beryllium,* ed. D. W. White, Jr. and J. E. Burke, (Cleveland: American Society of Metals, 1955), 1–20.

7. J. H. Sterner and M. Eisenbud, "Epidemiology of Beryllium Intoxication," *Archives of Industrial Hygiene and Occupational Medicine,* 4(1951): 123–51.

8. Eisenbud and Lisson, "Epidemiological Aspects."

9. M. Eisenbud and J. A. Quigley, "Industrial Hygiene of Uranium Processing," *Archives of Industrial Health,* 14(1956): 12–22.

10. Hanson Blatz and M. Eisenbud, "An Estimate of Cumulative Multiple Exposures to Radioactive Materials, Mallinckrodt Chemical Works, July 1942 to October 1949," (U.S. Atomic Energy Commission, New York Operations Office, November 20, 1950.)

5 AEC: STUDIES OF RADIOACTIVE FALLOUT

1. M. Eisenbud and J. H. Harley, "Radioactive Dust from Nuclear Detonations," *Science,* 117(1953): 141–47.

2. W. B. Harris, H. D. LeVine and M. Eisenbud, "Field Equipment for Collection and Evaluation of Toxic and Radioactive Contaminants," *Archives of Industrial Hygiene and Occupational Medicine,* 7(1953): 490–502; M. Eisenbud and J. H. Harley, "Radioactive Fallout in United States," *Science,* 121(1955): 677–80; M. Eisenbud, "Monitoring Network for Measuring Radioactive Fallout," *Journal of the American Water Works Association,* 48(1956): 659–64; M. Eisenbud and J. H. Harley, "Radioactive Fallout through September 1955," *Science,* 124(1956): 251–55; M. Eisenbud, "Global Distribution of Strontium-90 from Nuclear Detonations," *Scientific Monthly,* 84(1957): 237–44; M. Eisenbud and J. H. Harley, "Long-term Fallout," *Science,* 128(1958): 399–402; M. Eisenbud, "Deposition of Strontium-90 through October 1958," *Science* 130(1959): 76–80; idem, "Distribution of Radioactivity in Foods," *Federation Proceedings,* 22(1963): 1410–14.

3. *Worldwide Effects of Atomic Weapons,* (Santa Monica, California: RAND Corporation, 1953), (Rep. R-251-AEC).

4. R. Kobayashi and I. Nagai, "Cooperation by the United States in the Radiochemical Analysis," *Research in the Effects and Influences of the Nuclear*

Bomb Test Explosions, (Tokyo: Japan Society for the Promotion of Science, 1956), 1435–55.

5. U.S. Atomic Energy Commission, "News Release," (March 10, 1954).

6. U.S. Atomic Energy Commission, "Statement," (March 24, 1954).

7. U.S. Atomic Energy Commission, "Text of Statement by Chairman," (April 1, 1954).

6 AEC: FROM SCIENCE TO ADMINISTRATION

1. J. H. Harley, "Operation TROLL," (U.S. Atomic Energy Commission, 1956), (Report NYOO 4656).

2. L. R. Solon, W. M. Lowder, A. V. Zila, H. D. LeVine, H. Blatz, and M. Eisenbud, "Measurements of External Environmental Radiation in the United States," *Science,* 127(1958): 1183–84.

3. U.S. Atomic Energy Commission, "Minutes of the Advisory Committee for Biology and Medicine," (January 12–13, 1951).

4. Eisenbud and Quigley, "Industrial Hygiene of Uranium Processing."

7 NEW YORK UNIVERSITY MEDICAL CENTER

1. M. Eisenbud, "Foreword," *Radioactivity in Man,* ed. G. R. Meneely (Springfield, Illinois: C. C. Thomas & Company, 1961), xxi–xxiii; idem, "The Role of Whole Body Counters in the Evaluation of Hazards," *Radioactivity in Man,* 323–33.

2. G. R. Laurer and M. Eisenbud, "*In Vivo* Measurements of Nuclides Emitting Soft Penetrating Radiations," *Diagnosis and Treatment of Deposited Radionuclides Proceedings,* ed. H. A. Kornberg and W. D. Norwood, (New York: Excerpta Medica Foundation, 1968), 189–207.

3. M. M. Zaret and M. Eisenbud, "Preliminary Results of Studies of the Lenticular Effects of Microwaves among Exposed Personnel," *Biological Effects of Microwave Radiation,* ed. M. F. Peyton, (New York: Plenum, 1961), 293–308.

4. A. R. Sheppard and M. Eisenbud, *Biological Effects of Electric and Magnetic Fields of Extremely Low Frequency,* (New York: New York University Press, 1977).

5. R. B. Olcerst, S. Belman, M. Eisenbud, W. W. Mumford and J. R. Rabinowitz, "The Increased Passive Efflux of Sodium and Rubidium from Rabbit Erythrocytes by Microwave Radiation," *Radiation Research,* 82(1980): 244–56.

6. M. B. Heller, H. Blatz, B. Pasternack and M. Eisenbud, "Radiation Dose from Diagnostic Medical X-ray Procedures to the Population of New York City," *American Journal of Public Health,* 54(1964): 1551–59.

7. G. R. Laurer and M. Eisenbud, "Low-level *In Vivo* Measurement of Iodine-131 in Humans," *Health Physics,* 9(1963): 401–5; M. Eisenbud and M. E. Wrenn, "Biological Disposition of Radioiodine: A Review," *Health Physics,* 9(1963): 1133–39.

8. M. Blum and M. Eisenbud, "Reduction of Thyroid Irradiation from I-131 by Potassium Iodide," *Journal of the American Medical Association,* 200(1967): 1036–40; idem, "Therapeutic Reduction of Thyroidal Irradiation from I-131 by the Use of Potassium Iodide and Thyroid Stimulating Hormone," *1st Interna-*

tional Congress on Radiation Protection Proceedings, ed. W. S. Snyder et al., (New York: Pergamon Press, 1968): 1309–15.

9. M. Eisenbud, Y. Mochizuki, A. S. Goldin, G. R. Laurer, "Iodine-131 Dose from Soviet Nuclear Tests," *Science*, 136(1962): 370–374; M. E. Wrenn, R. Mowafy and G. R. Laurer, "Zr-95-Nb-95 in Human Lungs from Fallout," *Health Physics*, 10(1964): 1051–58; M. Eisenbud, Y. Mochizuki and G. R. Laurer, "I-131 Dose to Human Thyroids in New York City from Nuclear Tests in 1962," *Health Physics*, 9(1963): 1291–99; M. Eisenbud, B. Pasternack, G. Laurer and L. Block, "Variability of the I-131 Concentrations in the Milk Distribution System of a Large City," *Health Physics*, 9(1963): 1303–5; M. Eisenbud, B. Pasternack, G. Laurer, Y. Mochizuki, M. E. Wrenn, L. Block and R. Mowafy, "Estimation of the Distribution of Thyroid Doses in a Population Exposed to I-131 from Weapons Tests," *Health Physics*, 9(1963): 1281–89; M. Eisenbud, "Radioactive Fallout Problems in Food, Water, and Clothing," *Archives of Environmental Health*, 8(1964): 606–12; G. R. Laurer and M. Eisenbud, "Iodine Uptake Measurement With Nanocurie Doses of I-131," presented at *2d Symposium of Radioactivity in Man*, Chicago, Illinois, 1964; "Iodine-131 in Children's Thyroids from Environmental Exposure," G. J. Karches, H. N. Wellman, W. G. Hansen, G. R. Laurer, M. Eisenbud and J. H. Fooks, (July 1965), (Public Health Service Pub. no. 999-RH-14).

10. Peter Freudenthal, "Strontium-90 Concentrations in Surface Air: North America versus Atlantic Ocean from 1956 to 1969," *Journal of Geophysical Research*, 75(1970): 4089–96.

11. *International Symposium on Areas of High Natural Radioactivity Proceedings*, (Rio de Janeiro: Academia Brasileira de Ciências, 1977); E. Penna-Franca, J. C. Almeida, J. Becker, M. Emmerich, F. X. Roser and G. Kegel et al., "Status of Investigations in the Brazilian Areas of High Natural Radioactivity," *Health Physics*, 11(1965): 699–712.

12. M. A. Barcinski, M. C. A. Abreu, J. C. C. de Almeida, J. M. Naya, L. G. Fonseca and L. E. Castro, "Cytogenetic Investigation in a Brazilian Population Living in an Area of High Natural Radioactivity," *American Journal of Human Genetics*, 27(1975): 802–6.

13. M. Eisenbud, H. Petrow, R. T. Drew, F. X. Roser, G. Kegel and T. L. Cullen, "Naturally Occurring Radionuclides in Foods and Waters from the Brazilian Areas of High Radioactivity," *The Natural Radiation Environment*, ed. J. A. S. Adams and W. M. Lowder, (Chicago: University of Chicago Press, 1964), 837–54; M. Eisenbud, "The Natural Versus the Unnatural," *Second International Symposium on Natural Radiation Environment Proceedings* (Washington, D.C.: Energy Research and Development Administration, CONF 720805-P2, 1972, 941–947; R. T. Drew and M. Eisenbud, "The Natural Radiation Dose to Indigenous Rodents on the Morro do Ferro, Brazil," *Health Physics*, 12(1966): 1267–74; idem, "The Pulmonary Dose from Rn-220 Received by Indigenous Rodents of the Morro do Ferro, Brazil," *Radiation Research*, 42(1970): 270–81; M. Eisenbud, "Summary Report," *International Symposium on Areas of High Natural Radioactivity Proceedings* (Rio de Janeiro: Academia Brasileira de Ciências, 1977), 167–79.

14. M. Eisenbud, W. Lei, R. Ballad, K. Krauskopf, E. Penna-Franca and T. L. Cullen et al., "Mobility of Thorium from the Morro do Ferro," *Environmental Migration of Long-Lived Radionuclides*, (Vienna: International Atomic Energy Agency, 1982), 739–55; M. Eisenbud, W. Lei, R. Ballad, E. Penna-Franca, N. Miekeley and T. Cullen et al., "Studies of the Mobilization of Thorium from the Morro do Ferro," *Scientific Basis for Nuclear Waste Management V*, ed. W. Lutze (New York: North-Holland/Elsevier Science Publishing Company, 1982), 735–44; M. Eisenbud, K. Krauskopf, E. Penna-Franca, W. Lei, R. Ballad and P. Linsalata et al, "Natural Analogues for the Transuranic Actinide Elements: An Investigation in Minas Gerais, Brazil," *Environmental Geology and Water Sciences*, 6(1984): 1–9; P. Linsalata, M. Eisenbud and E. Penna-Franca, "Thorium and the Light Rare Earth Elements in Soil, Crops and Domestic Animals from Abnormally High and Typical Radiation Background Areas," *Environmental Radiation '85 (Proceedings of the 18th Midyear Topical Symposium of the Health Physics Society)*, (Laramie, Wyoming: Central Rocky Mountain Chapter, Health Physics Society, 1985), 535–42; W. Lei, P. Linsalata, E. Penna-Franca and M. Eisenbud, "Distribution and Mobilization of Cerium, Lanthanum and Neodymium in the Morro do Ferro Basin, Brazil," *Chemical Geology*, 55(1986): 313–22.

15. M. Eisenbud, G. R. Laurer, J. C. Rosen, N. Cohen and J. Thomas, "*In Vivo* Measurement of Lead-210 as an Indicator of Cumulative Radon Daughter Exposure in uranium Miners," *Health Physics*, 166(1969): 637–46; M. E. Wrenn, N. Cohen, J. C. Rosen, M. Eisenbud, "*In Vivo* Measurements of Lead-210 in Man," *Assessment of Radioactive Contamination in Man*, (Vienna: International Atomic Energy Agency, 1972), 129–46;

16. M. Eisenbud and G. Gleason, eds., *Electric Power and Thermal Discharges*, (New York: Gordon and Breach, 1969).

17. G. P. Howells, T. J. Kneip and M. Eisenbud, "Water Quality in Industrial Areas: Profile of a River," *Environmental Science Technology*, 4(1970): 26–35; J. Lentsch, M. E. Wrenn, T. Kneip and M. Eisenbud, "Manmade Radionuclides in the Hudson River Estuary," *Proceedings—5th Annual Midyear Topical Symposium* (at Idaho Falls, ID, Nov. 1970), (Health Physics Society, 1970): 499–528; M. E. Wrenn, J. W. Lentsch, M. Eisenbud, G. J. Laurer and G. P. Howells, "Radiocesium Distribution in Water, Sediment, and Biota in the Hudson River Estuary from 1964 through 1970," *Proceedings—3d National Symposium on Radioecology*, (Washington, D.C.: US AEC Conf-710501-P1, 1971), 334–43; J. W. Lentsch, T. J. Kneip, M. E. Wrenn, G. P. Howells and M. Eisenbud, "Stable Manganese and Mn-54 Distributions in the Physical and Biological Components of the Hudson River Estuary," *Proceedings—3d National Symposium on Radioecology*, (Washington, D.C. US AEC Conf-710501-P2, 1971), 752–68; G. P. Howells, M. Eisenbud and T. J. Kneip, "Ecology of the Estuary of the Lower Hudson River," *Proceedings—FAO Technical Conference on Marine Pollution and Its Effects on Living Resources and Fishing*, (New York: Food and Agriculture Organization of the United Nations, Pub no. FIR:MP/70/E-18 1970), 1–7; M. Eisenbud, "Hudson River Ecology in Historical Perspective," *Proceedings—2d Symposium on Hudson River Ecology* (at Tuxedo, New York,

sponsored by New York University), (Albany, New York: New York State Department of Environment and Conservation, 1969), 11–19; A. S. Paschoa, M. E. Wrenn and M. Eisenbud, "Natural Radiation Dose to *Gammarus* from Hudson River," *Proceedings—4th Symposium on Hudson River Ecology* (Bear Mountain, New York, Nov. 1976), (Bronx, New York: Hudson River Environmental Society, Paper no. 11, 1976), 1–31; idem, Natural Radiation Dose to *Gammarus* from Hudson River," *Radioprotection DUNOD*, 14(1979): 99–115.

18. M. Eisenbud and H. Petrow, "Radioactivity in the Atmospheric Effluents of Nuclear Power Plants That Use Fossil Fuels," *Science*, 144(1964): 288–89.

8 ENVIRONMENTAL PROTECTION ADMINISTRATOR, CITY OF NEW YORK

1. T. J. Kneip, M. Eisenbud, C. D. Strehlow and P. C. Freudenthal, "Airborne Particulates in New York City," *Journal of the American Public Health Association*, 20(1970): 144–9; M. Eisenbud and L. Ehrlich, "Carbon Monoxide Concentration Trends in Urban Atmospheres," *Science*, 176(1972): 193–4; M. Eisenbud and T. J. Kneip, "Trace Metals in the Atmosphere," (July 1971), (New York State Department of the Environment and Conservation, Technical Paper no. 16); M. T. Kleinman, T. J. Kneip and M. Eisenbud, "Meteorological Influences on Airborne Trace Metals and Suspended Particulates," *Trace Substances in Environmental Health VIII*, ed. D. D. Hemphill (Columbia, Missouri: University of Missouri Press, 1974), 147–59; D. M. Bernstein, T. J. Kneip, M. T. Kleinman, R. Riddick and M. Eisenbud, "Uptake and Distribution of Airborne Trace Metals in Man," *Trace Substances in Environmental Health VIII*, ed. D. D. Hemphill (Columbia, Missouri: University of Missouri, 1974), 329–34; M. T. Kleinman, T. J. Kneip and M. Eisenbud, "Seasonal Patterns of Air-borne Particulate Concentrations in New York City," *Atmospheric Environment*, 10(1976): 9–11; T. J. Kneip, M. Lippman, M. Eisenbud, D. M. Bernstein and M. T. Kleinman, "Trace Elements in the Urban Atmosphere and Human Respiratory Tissue," *Proceedings American Petroleum Institute 44th Midyear Meeting*, Refining Department, 58(1979): 355–61; M. T. Kleinman, B. Pasternack, M. Eisenbud and T. J. Kneip, "Identifying and Estimating the Relative Importance of Sources of Air-borne Particulates," *Environmental Science Technology*, 14(1980): 62–65; M. Morandi and M. Eisenbud, "Carbon Monoxide Exposure in New York City: A Historical Overview," *Bulletin of the New York Academy of Medicine*, 56(1980): 817–28.

2. M. Eisenbud, "Environmental Protection in the City of New York," *Science*, 170(1970): 706–12.

9 MEETING, WRITING, AND ADVISING

1. M. Eisenbud, "The Principal Health Hazards."

2. M. Eisenbud, "Fallout: Fact and Opinion," *New York Times Magazine*, (January 10, 1960): 77–81; idem, "Educating the Public about Radiation (guest editorial)," *Nucleonics*, (1960); idem, "The Risks," *America Faces the Nuclear Age*, (New York: Sheridan House, 1961), 91–106; idem, "Environmental Safety in the Nuclear Age," *Energy, Public Policy and the Law*, ed. E. J. Bloustein

(Dobb's Ferry, New York: Oceana Publications for New York University School of Law, 1964), 51–58; idem, "Industrial Uses of Ionizing Radiation," *American Journal of Public Health,* 55(1965): 748–59; idem, "Radiation in Perspective," *Nuclear Safety,* 6(1964): 380–85; idem, "Environmental Pollution and Its Control," *Bulletin of the New York Academy of Medicine,* 45(1969): 447–54; idem, "Standards of Radiation Protection and Their Implications for the Public's Health," *Nuclear Power and the Public,* ed. H. Foreman (Minneapolis: University of Minnesota Press, 1970), 73–87; idem, "Technology and Man," *Proceedings Anglo-American Conference on Environmental Control,* (Oxfordshire, England: The Ditchley Foundation, 1970), 16–28; idem, "Environmental Controls—Political Or Analytical?," *Society for Industrial and Applied Mathematics (SIAM) News,* 66(1973): 2–3; idem, "Environmental Protection in Historical Perspective," *Proceedings—2d Annual Israel Conference on Environmental Quality,* (Jerusalem: Environmental Protection Service, Prime Minister's Office, May 1973), 4–13; idem, "Contributions of Electric Power for Environmental Betterment," *Proceedings—Conference on Research for the Electric Power Industry,* (New York: Institute of Electrical & Electronics Engineers, 72 CHO 726-0 PWR, 1973), 391–94; idem, "The Natural vs. the Unnatural," *Proceedings—2d International Symposium on Natural Radiation Environment II,* (Washington, D.C.: US ERDA Conf-720805-P2, 1972), 941–47; idem, "Environmental Causes of Cancer," *Environment,* 20(1978); 6–16; idem, "Medicine and Public Health," *Science, Technology, and the Human Prospect: Proceedings of the Edison Centennial Symposium,* ed. C. Starr and P. C. Ritterbush, (New York: Pergamon Press, 1980), 169–76; idem, "Radioactive Wastes from Biomedical Institutions (editorial)," *Science,* 207(1980): 1299; idem, "The Environment, Technology and Health: A Century of Progress but a Time of Despair," *American Journal of Medicine,* 68(1980): 476–78; idem, "The Concept of *de Minimis* Dose," *Quantitative Risk in Standard Setting,* (Washington, D.C.: National Council on Radiation Protection & Measurements, 1981), 64–75; idem, "The Human Environment—Past, Present and Future," *Environmental Radioactivity Proceedings No. 5,* (Washington, D.C.: National Council on Radiation Protection and Measurements, 1983), 201–28.

3. M. Eisenbud, *Environmental Radioactivity,* 3d ed.

4. M. Eisenbud, *Environment, Technology, and Health,* (New York: New York University Press, 1978).

5. M. Eisenbud, "Low Level Radioactive Waste Repositories: A Risk Assessment," *Journal of the Elisha Mitchell Scientific Society,* 104(1989): 76–80; idem, "Management Strategies for Low Level Radioactive Waste Disposal," *Bulletin of the New York Academy of Medicine,* 65(1989): 451–500.

10 FROM THE PAST TO THE PRESENT

1. M. Eisenbud, "Technology and Man."

2. M. Eisenbud, "Environmental Causes of Cancer."

3. M. Eisenbud, *Environment, Technology, and Health.*

4. M. Eisenbud, "Disparate Costs of Risk Avoidance," *Science,* 241(1988).

5. M. Eisenbud, "Exposure of the General Public Near Three Mile Island," *Nuclear Technology* 87 (1989):514–19

6. R. Wilson, "A Visit to Chernobyl," *Science* 236(1987): 1635–40.

7. M. Eisenbud, "Exposure of the General Public Near Three Mile Island."

INDEX